A CONTEXTUAL APPROACH TO INTEGRATED MATHEMATICS

Unit 9

Using Ratios and Proportions

Teacher's Guide

Developed by the Center for Occupational Research and Development and sponsored by a consortium of State Vocational Education Agencies with the cooperation and support of mathematics educators.

©Copyright 1988 by the Center for Occupational Research and Development
Waco, Texas

Neither CORD nor any member of the consortium assumes any liabilities with respect to the use of, or for damages resulting from the use of, any information, apparatus, method or process described in these materials.

Published and distributed by: CORD Communications
324 Kelly Drive, Waco, Texas 76710
817-776-1822 Fax 817-776-3906

Printed in USA, September, 1994

ISBN 1-55502-296-0 (Applied Mathematics)
ISBN 1-55502-310-X (Unit 9, Using Ratios and Proportions, Teacher's Guide)

PREFACE TO THE TEACHER'S GUIDE

The Teacher's Guide of **Applied Mathematics** contains information that will help you present the subject matter found in the student text. The Teacher's Guide, located in the front half of this unit, consists of three main parts: (1) introductory information; (2) how to teach the unit and (3) teacher resource materials. The student materials, located in the back half of this unit, are identical to those found in the corresponding unit of the student text, with the important exception that selected pages are annotated to support the teaching process. Some annotations supplement, expand or otherwise interpret the information in the text; others suggest alternate techniques for presentation of the material.

In the Teacher's Guide, the first section, *Introduction*, provides you with insight into the philosophy that forms the basis for **Applied Mathematics** as well as special features of the student material and suggested strategies for teaching the materials. The second section, *Teaching the Unit*, provides a unit breakdown of the separate parts to be covered and a suggested teaching path for each part. The third section, *Resources*, contains (1) helpful information about the unit video, (2) equipment lists and appropriate notes relating to the hands-on laboratory activities, (3) complete solutions to all of the problem-solving exercises and (4) a problem bank for end-of-unit test with answers. In addition, for certain units, this section will contain optional "drill" exercises for students who may need more practice. Finally, for the three preparatory units (A, B and C) and the first regular unit, diagnostic tests will be provided to help you evaluate the level of student preparation for the **Applied Mathematics** materials.

Table of Contents

PART 1: TEACHER INFORMATION

Introduction T-1
- Features of Applied Mathematics T-2
- Teaching Strategy T-3

Teaching the Unit T-5
- Teaching Path – Session 1 ..... T-6
- Teaching Path – Sessions 2 and 3 ..... T-7
- Teaching Path – Session 4 ..... T-9
- Teaching Path – Session 5 ..... T-10
- Teaching Path – Session 6 ..... T-11

Resources T-12
- About the Video, "Using Ratios and Proportions" T-13
- Laboratory Activities T-15
- Solutions to Student Exercises T-21
- Student Handouts T-43
- Optional Skill Drill T-50
- Problem Bank for End-of-Unit Test T-55

PART 2: STUDENT TEXT

Learning the Skills
- Introduction 3
- Identifying Ratios 4
- Reading and Interpreting Ratios 8
- Comparing Ratios 10
- Identifying and Writing Proportions 14
- Solving Proportions 16
- Summary 27

Practicing the Skills
- Laboratory Activities 29
- Student Exercises 34

Reference Materials
- Glossary 51

PART 1: TEACHER INFORMATION

INTRODUCTION

Applied Mathematics is a set of modular learning materials prepared to help high school vocational students and others develop and refine job-related mathematics skills. The overall course includes material that focuses on arithmetic operations, problem-solving techniques, estimation of answers, measurement skills, geometry, data handling, simple statistics, and the use of algebraic formulas to solve problems. Though the text includes some material found in traditional areas of arithmetic, geometry, algebra, and simple trigonometry, the emphasis remains on the ability to understand and apply functional mathematics to solve problems in the world of work.

The student materials are designed to be used in a one-year course for academic credit toward high school graduation. Or, alternatively, they may be used in part and infused, as needed, into existing vocational courses. They have been prepared for learners who have an eighth-grade, pre-algebra mathematics competency level and who may feel a certain anxiety about mathematics. They are written generally at an eighth-grade reading level.

The materials are deemed appropriate for high school vocational students in grades nine through twelve, and for young persons and adults, out of high school, who are engaged in vocational and technical training. The post-high school learners may be found in skill centers, Job Corps centers or in learning centers involved in preemployment training, retraining, upgrade training, or apprenticeship training.

Applied Mathematics may be taught by teachers who understand the basic concepts of mathematics and who can communicate to students how mathematics is used in today's workplace. If the materials are taught by certified teachers of mathematics, academic credit for mathematics can be expected.

FEATURES OF APPLIED MATHEMATICS

Applied Mathematics is made up of 22 separate, modular units. They are
1. Learning Problem-solving Techniques
2. Estimating Answers
3. Measuring in English and Metric Units
4. Using Graphs, Charts, and Tables
5. Dealing with Data
6. Working with Lines and Angles
7. Working with Shapes in Two Dimensions
8. Working with Shapes in Three Dimensions
9. Using Ratios and Proportions
10. Working with Scale Drawings
11. Using Signed Numbers and Vectors
12. Using Scientific Notation
13. Precision, Accuracy, and Tolerance
14. Solving Practical Problems with Powers and Roots
15. Using Formulas to Solve Practical Problems
16. Solving Word Problems That Involve Linear Equations
17. Graphing Data
18. Solving Word Problems That Involve Nonlinear Equations
19. Working with Statistics
20. Working with Probabilities
21. Using Right-triangle Relationships
22. Using Trigonometric Functions

In addition, for those learners who may not have an adequate working knowledge of eighth-grade pre-algebra mathematics, three optional review units have been prepared. They are entitled:
A. Getting to Know Your Calculator
B. Naming Numbers in Different Ways
C. Finding Answers with Your Calculator

Each of these units encourages the students to use calculators while they review the main ideas that deal with numbers, fractions, decimals, and percents.

The **Applied Mathematics** learning materials may be presented as a stand-alone course, beginning with the three optional units (if needed) and moving sequentially through the remaining 22 units. Or, a certain subsection of the 25 units may be selected and used as appropriate mathematics materials in support of specific vocational studies in such areas as health occupations, automotive technology, home economics or agriculture.

Each unit in the **Applied Mathematics** learning materials is an integrated learning package made up of supporting parts: video program, text, mathematics laboratory activities, practical problem-solving exercises, and glossary. Audio-visual support is provided by a unit video program. This program introduces the mathematics unit and sets the stage for the relevance of mathematics in the world of work. The mathematics concepts for each unit are explained carefully in the student text. The reader is invited to play a participatory role in the learning process. Participation is encouraged by activities the learner performs while studying the material. Hands-on mathematics activity laboratories are provided to make the mathematics concepts "come to life" and be more useful. The activities are performed in the classroom or school environment (generally) with relatively inexpensive equipment. Finally, a set of 40 problems that apply the unit mathematics concepts to general and occupationally specific practical situations is made available for use as needed. These problems are almost always "word problems" and range in difficulty from the less difficult to the more difficult.

TEACHING STRATEGY

Each of the 25 units of **Applied Mathematics** may be taught in a similar manner. The teaching strategy outlined below is meant to serve only as a GUIDE—as a SUGGESTED teaching pattern. Different teachers, in different learning environments, with different student populations necessarily will modify the pattern to suit their individual needs. By design, the learning materials have been prepared to offer the teacher sufficient flexibility to expand or contract the time devoted to each unit. For example, teachers may choose to involve students in one, two or all of the suggested laboratory activities. In the same spirit, teachers may choose to assign 5, 10, 15 or more of the 40 problems provided in each unit.

The suggested teaching plan is divided into six activity sessions. The sessions concentrate on "Getting Started," "Learning the Skills," "Applying the Skills" and "Looking Back." It is important to note that we are NOT equating six sessions with six traditional class periods of 40 or 50 minutes each. Rather, we suggest that the six sessions serve as a guide to the intended division of activities, and that the length of time devoted to each session—based on student backgrounds, course goals and time available—be determined by the teacher.

A Flexible Teaching Plan

Session 1: Getting Started	Sessions 2 and 3: Learning the Skills	Sessions 4 and 5: Applying the Skills	Session 6: Looking Back
• Overview of Unit • Video • Class Problem	• Read and Do • Class Activities • Class Discussions • Study Examples	• Hands-on Mathematics Lab • Solve Problems • Use Calculators	• Review • Evaluate Programs

<u>Session 1</u> is introductory in nature. It involves overviewing the unit, viewing and discussing the video, and working—as a class—the problem proposed in the video. Session 1 sets the stage for the mathematics skills to be learned.

<u>Sessions 2 and 3</u> deal with the mathematics concepts presented in the unit. Through the "read and do" material in the text, class activities, class discussion and study of examples, the students learn about the basic skills presented in the unit.

<u>Sessions 4 and 5</u> focus on applying the mathematics skills in lab-oriented activities and problem-solving exercises. The use of calculators is strongly encouraged throughout these activities. Students complete the lab activities as group efforts, collecting data, making calculations and discussing their results. Similarly, students work the problem-solving exercises in groups—in an informal atmosphere—helping one another, discussing approaches and solutions with each other and with the teacher, as required.

<u>Session 6</u> is a wrap-up session. It is appropriate to review the unit objectives, watch the video again, and recap the ideas presented in the summary. An end-of-unit test to evaluate student progress may be administered at this time.

TEACHING THE UNIT

The "unit breakdown" that follows provides you with information about the different learning materials contained in this unit. Class activities are keyed to the sessions described in the aforementioned "flexible teaching plan" and appropriate page numbers are listed for the main parts of the student text.

Breakdown of Unit 9: "Using Ratios and Proportions"

Session	Student Text Page #	Class Activity
1	1-4	Introduce the unit Watch unit video (8 min) Solve "video" class activity
2	4-16	Discuss reading assignments Conduct class activities
3	16-28	Work through examples
4	29-33	Participate in lab activities
5	34-50	Practice problem solving
6	27-28	Review the unit Give a unit test

Video playing time – 8:18 minutes

TEACHING PATH – SESSION 1

RESOURCE MATERIALS

Video: "Using Ratios and Proportions" (time – 8:18 minutes)
About the Video: See Resource pages
Student Text: "Using Ratios and Proportions"

SESSION GOALS

This session introduces students to ratios and proportions. Have your students discuss common uses of ratios. Encourage them to distinguish between ratios and proportions. Let them describe how they have used ratios and proportions to solve everyday problems.

CLASS ACTIVITIES

1. Read over the **Unit Objectives** with students. Discuss them briefly.

2. Summarize the material included in the **Introduction**. Be sure to read the annotated teacher comments that accompany the student pages.

3. Show the video, "Using Ratios and Proportions." It is about eight minutes in length. Refer to the part entitled "About the Video" in the *Resources* section of this guide. There you will find suggested lead-in questions to be used before you turn on the video, a summary of what the video is about, and a restatement of the suggested class problem posed by the video. (We also have provided you with a solution to the "video problem.") After showing the video, ask students to recall situations in the video where the role models used ratios and proportions to solve problems. How many similar applications can the students think of from other occupations?

4. Use the problem posed in the video as an activity to be solved in class. You can lead the solution by doing it on the chalkboard—or you can divide the class into teams to solve the problem.

5. Assign the text material for Session 2 to be read before coming to class. (See "Breakdown" table.)

TEACHING PATH – SESSIONS 2 AND 3

RESOURCE MATERIALS
Student Text:
> Read up to "Solving Proportions" for Session 2.
> Read through the "Summary" for Session 3.

SESSION GOALS

Learn the mathematics skills covered in Objectives 1-5.

CLASS ACTIVITIES: SESSION 2

1. Summarize the main points of text material assigned for Session 2. Also, see the appropriate annotated pages in the student-text portion of the Teacher's Guide.
2. Work through some (or all) of the study activities presented in the text. Have your students practice setting up ratios, interpreting ratios and comparing ratios. Discuss similar figures. Have students determine if two ratios are proportional or not.
3. Assign reading homework for Session 3. (See "Breakdown" table.)

CLASS ACTIVITIES: SESSION 3

1. Summarize the main points of the text material assigned for Session 3. See annotated student pages.
2. Work through some (or all) of the activities and examples presented in the text. Have students practice setting up and solving proportions. Let them discuss what kinds of problems can be solved by using proportions. Help them to understand direct and inverse proportions and how to recognize each type.
3. Have students read through the **Lab Activity Problem(s)** you have selected for the next session, Session 4, **before** they come to the session.
4. Assign 5 **general** problems and 5 (or more) of the 35 **occupationally related** problems. To help you select problems, consult the "problem" matrix that follows this page. It provides you with a breakdown of the 40 problems according to areas of application (general; agriculture and agribusiness; business and marketing; health occupations; home economics; and industrial technology). A letter in a given block indicates both the area of applicability and the level of difficulty of the problem. The level of difficulty ranges from A (least difficult) to C (most difficult). Ask students to begin solving problems before Session 5. Tell them that they will complete the problem-solving activity in Session 5.

STUDENT EXERCISES FOR UNIT 9
AREA OF APPLICATION AND LEVEL OF DIFFICULTY

Exer. No.	General	Agriculture Agribusiness	Business & Marketing	Health Occupations	Home Economics	Industrial Technology
1	A					
2	A					
3	B					B
4	C					
5	C					
6		B				
7		A	A			
8		B				B
9	B	B				
10		C				
11			A			
12		A	A			
13			A			
14		B	B			
15			B			B
16			C			
17	C		C			
18			C			
19			C		C	
20			C			
21				A		
22				A		
23				B		
24			C	C		
25				C		C
26			B		B	
27					B	
28				C	C	
29	C	C			C	
30	C		C		C	
31		B				B
32		B				B
33						B
34		B				B
35						B
36						B
37		C				C
38						C
39						C
40						C

KEY TO CODES
A: Less Difficult
B: Average Difficulty
C: More Difficult

TEACHING PATH – SESSION 4

Note: This session requires students to use measuring apparatus, collect data, and solve a meaningful problem in a laboratory situation. There are three suggested lab activities. Select one or more—as time permits—to be completed by your students. Divide your class into teams of three or four students. Assign activities. ENCOURAGE STUDENTS TO USE CALCULATORS WHENEVER COMPUTATIONS ARE TO BE MADE.

RESOURCE MATERIALS
Student Text:
    Lab Activity Problems
        Activity 1 – Ratios and proportions in similar triangles
        Activity 2 – Using ratios to make scale drawings
        Activity 3 – Lever arm and force ratios

Teacher's Guide:
    Notes on laboratory activities. See *Resources*.

CLASS GOALS

Involve students in using problem-solving skills.

CLASS ACTIVITIES

1. Have students read through each laboratory activity you have assigned. Ask them to pay special attention to what is given and what is to be found.

2. Have students carry out the recommended Procedures for each laboratory activity assigned. Provide help as needed, but encourage students to be creative and to use their own ideas. Be sure students understand the problem(s) to be solved.

3. Have a member of each team report the results of the team's finding—to you or to the class as a whole—as time permits.

TEACHING PATH – SESSION 5

According to the suggested plan for teaching this unit, you were to have assigned selected problems to be solved at the end of Session 3. Presumably, then, you assigned five **general** problems and five or more **occupationally specific** problems. Your students have been working on these as homework problems. In this class, Session 5, **problem solving** is to continue, under your supervision and with your help.

RESOURCE MATERIALS
Student Text:
   Problem Bank of General and Occupationally Specific Problems
Teacher's Guide:
   Solutions to Student Exercises. See *Resources*.

CLASS GOALS
Provide students with practice in solving problems involving ratios and proportions.

CLASS ACTIVITIES
1. Continue solving the problems you assigned in Session 3. Make class an informal problem-solving activity. Help students as needed. Encourage students to help one another.
2. Encourage students to use calculators as needed.
3. Allow time to discuss problems that may be giving students some difficulty.
4. Tell students that the next session (Session 6) is the last session of the unit. Ask them to review the **Unit Objectives** and reread the **Unit Summary**. Tell them that after a short review they will be given an end-of-unit test.

TEACHING PATH – SESSION 6

This session is intended for review and evaluation. You may want to tie up any loose ends that weren't completed during the previous sessions. Be sure to leave enough time for review and testing.

RESOURCE MATERIALS
Video: "Using Ratios and Proportions"
Student Text: Unit Objectives/Summary/Glossary
Unit Test: See *Resources* section.

CLASS ACTIVITIES
1. Show the video "Using Ratios and Proportions" again. Students will understand the contents of the video much better the second time around.

2. Review the **Unit Objectives** and use the **key icons** and **Unit Summary** to recap the important mathematics skills presented in the unit. Go over the words in the **Glossary** to help students focus on key words and key ideas.

3. Give the **Unit Test**.

4. Tell students which unit they will be covering next. Ask them to read the **Unit Objectives** and **Introduction** for that unit before coming to the next class.

RESOURCES

This section contains the following resources to help you teach this unit:

- **ABOUT THE VIDEO**—This resource provides you with a summary of what the video program is about, suggested lead-in questions to use with your students before you show the video, and a restatement of the "class problem" posed in the video program. A suggested solution to the class problem is included.

- **LABORATORY ACTIVITIES**—This resource provides you with a list of equipment required for each of the laboratory activities and an approximate cost. In addition, helpful notes on the laboratory activities, with typically expected outcomes, are included.

- **SOLUTIONS TO PROBLEM-SOLVING EXERCISES**—This resource provides you with detailed solutions to each of the 40 general and occupationally specific problems found in the student text.

- **STUDENT HANDOUTS**—This resource includes instructions for using the Accu-Line™ drawing kit and tables of conversion factors—resources students may need to complete the laboratory activities and student exercises.

- **OPTIONAL SKILL DRILL**—This optional resource provides you with a problem bank of routine practice exercises for those students who may need more help. Answers are provided so that students can check their progress and obtain immediate reinforcement for correct work.

- **END-OF-UNIT TEST**—This resource provides you with a test to measure student understanding of the mathematics concepts and applications covered in the unit. The test, usually made up of 30 questions, generally includes several multiple-choice entries, word problems that require calculations, sketches and graphs to be drawn and interpreted, and mathematics-skills activities to be performed.

ABOUT THE VIDEO, "USING RATIOS AND PROPORTIONS"

Before the Video

Ask students the following questions:

- What is a ratio? (A numerical comparison of two quantities.)
- What is a proportion? (An equality between two ratios.)
- What are some typical applications of ratios and proportions? (Increasing recipes, mixing solutions, and establishing seed and fertilizer needs.)

Summary of the Video

This video begins with a bicycle trick. The cyclist states that he can do the trick only in third gear, because that's the only gear that has the right gear ratio. Explaining what he means by "gear ratio," he says that ratio and proportion apply to bicycles, cars, cooking, and "you name it."

The video then shows Chef Andrew Cardy modifying a recipe by adjusting the ratios of ingredients proportionately to double the number of servings. Doubling the ratio of flour to butter (1.5 : 1) gives a ratio of 3 : 2. The narrator explains that ratios can be multiplied or divided, as long as you do it to both parts of the ratio.

The next situation shows a farmer preparing a feed mixture for hens of 4 parts corn to 1 part soybeans. This ratio gives the correct percent of protein for hens. Other animals, he explains, will have requirements for different ratios.

Ratios are shown again as a farmer makes plans for planting by using the ratio of 100 pounds of seed per acre (100 : 1). To figure how much seed is needed for 4 acres, the farmer uses a proportion, which is explained as "an equality between two ratios:"

$$\frac{100 \text{ (lb seed)}}{1.0 \text{ (acre)}} = \frac{400 \text{ (lb seed)}}{4.0 \text{ (acres)}}$$

The concept of ratio and proportion, the narrator explains, has numerous agricultural, industrial, and commercial applications. Often, the ratios express relationships among many different ingredients.

A final example shows an aerial photographer scaling a photograph in the ratio of 1 : 100.

Restatement of the Video Problem

The video asks the students to calculate the amount of water that should be mixed with 75 gallons of insecticide concentrate if the ratio is to be 3 parts of insecticide to 5 parts of water.

Suggested Solution to the Video Problem

To solve the problem, set up and solve the proportion:

$$\frac{3 \text{ parts (insecticide)}}{5 \text{ parts (water)}} = \frac{75 \text{ gallons (insecticide)}}{y \text{ gallons (water)}}$$

Solving for the unknown y gives

$$y = \frac{75 \text{ gallons} \times 5 \text{ parts}}{3 \text{ parts}}$$

$$y = 125 \text{ gallons of water}$$

LABORATORY ACTIVITIES

The first part of this resource for Laboratory Activities itemizes the materials (specific items of supply and equipment) required for each of the proposed activities. An estimated cost of materials for a class of 20-25 students—assuming 4 to 5 students per activity group—is included for each activity. The second part of this resource provides you with notes and expected outcomes for each of the proposed activities.

Materials Needed

The estimated costs are approximate and will vary from school to school. The costs will be less if you already have some of the materials. Note also that some of the materials cited here may be used in laboratory activities in other units. Also note that calculators are listed—for sake of completeness—even though they are not included in the cost of the materials for the specific activity.

Activity 1: Ratios and proportions in similar triangles
- Tape measure
- String
- Masking tape
- Carpenter's square
- Two ¼-inch dowel rods of lengths 18 inches and 36 inches
- Calculator

Estimated cost $116.00

Activity 2: Using ratios to make scale drawings
- Drawing kit (Accu-Line™)
- Tape measure
- Ruler
- Calculator

Estimated cost $330.00

Activity 3: Lever arm and force ratios
- Meter stick
- Spring scales, 5000-gram capacity, two
- Tape measure
- Fulcrum
- Calculator

Estimated cost $150.00

Teacher Notes: Laboratory Activities

ACTIVITY 1: Ratios in similar triangles

Procedure

a. The length along the floor is 8 feet. This is a 3-4-5 triangle.

b. and c. The results for Procedures b and c are shown in the following data table.

Length of dowel	Length along floor	Length along string
18 inches	24 inches	30 inches
36 inches	48 inches	60 inches
(wall distance is 72 inches)	(floor distance is 96 inches)	(string length is 120 inches)

Calculations

a. $\dfrac{\text{Length of rod (DE)}}{\text{Length of string segment (DC)}} \stackrel{?}{=} \dfrac{18 \text{ in.}}{30 \text{ in.}} \stackrel{?}{=} \dfrac{36 \text{ in.}}{60 \text{ in.}}$

b. Yes, the ratios are each equal to $3/5$.

c. $\dfrac{\text{Length of rod (DE)}}{\text{Distance on floor (CE)}} \stackrel{?}{=} \dfrac{18 \text{ in.}}{24 \text{ in.}} \stackrel{?}{=} \dfrac{36 \text{ in.}}{48 \text{ in.}}$

d. Yes, the ratios are each equal to $3/4$.

e. $\dfrac{\text{Length of string segment (DC)}}{\text{Distance on floor (CE)}} \stackrel{?}{=} \dfrac{30 \text{ in.}}{24 \text{ in.}} \stackrel{?}{=} \dfrac{60 \text{ in.}}{48 \text{ in.}}$

f. Yes, the ratios are each equal to $5/4$.

g. The ratio of $\frac{\text{wall (AB)}}{\text{string (AC)}}$ is $3/5$; it is equal to the ratio of $\frac{\text{rod (DE)}}{\text{string segment (DC)}}$.

The ratio of $\frac{\text{wall (AB)}}{\text{floor (BC)}}$ is $3/4$; it is equal to the ratio of $\frac{\text{rod (DE)}}{\text{floor segment (EC)}}$.

The ratio of $\frac{\text{string (AC)}}{\text{floor (BC)}}$ is $5/4$; it is equal to the ratio of $\frac{\text{string segment (DC)}}{\text{floor segment (EC)}}$.

For any one dowel rod, the following proportions are true:

rod : wall = string segment : string

rod : wall = floor segment : floor

string segment : string = floor segment : floor

These three proportions indicate that the three ratios—**rod : wall, string segment : string,** and **floor segment : floor**—are equal for **either** of the dowel rods.

Also, for the pair of dowel rods, the following proportions can be written:

rod$_1$: rod$_2$ = string segment$_1$: string segment$_2$

rod$_1$: rod$_2$ = floor segment$_1$: floor segment$_2$

string segment$_1$: string segment$_2$ = floor segment$_1$: floor segment$_2$

ACTIVITY 2: Using ratios to make scale drawings
Procedure

The following data are given for a "typical" classroom.

a. The length of the east or west classroom wall is 38 feet 9 inches. The width of north or south classroom wall is 20 feet 3 inches.

b. The length of the teacher's desk is 59 $3/4$ inches. The width of the teacher's desk is 30 inches. The teacher's desk is located in the southeast part of the room. The southeast corner of the desk is 3 feet 9 inches from the south wall and 5 feet from the east wall.

c. There are six windows in the south wall. Each one measures 3 feet by 5 feet. The height of the south wall is 8 feet 6 inches. The top of each window is 1 foot 3 inches from the ceiling, the east edge of the east window is 1 foot 3 inches from the east wall, and the windows are 1 foot 3 inches from each other.

Calculations

a. Drawn to scale 1" = 4'. Photographically reduced to 80% of full scale.

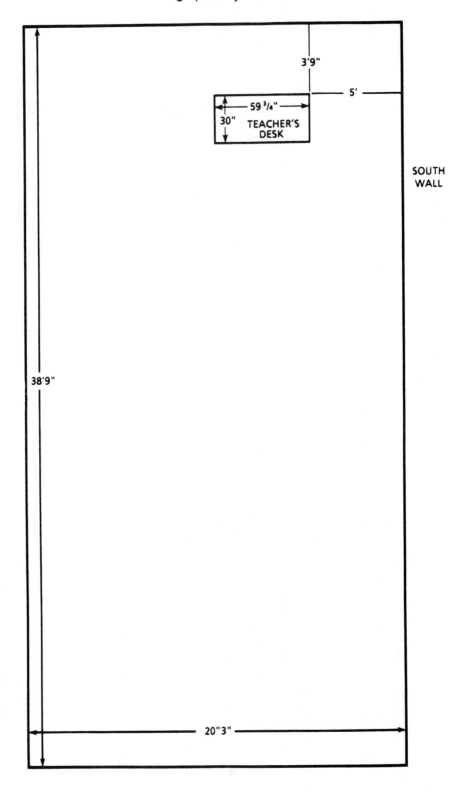

b. Drawn to scale 1" = 4'. Photographically reduced to 80% of full scale.

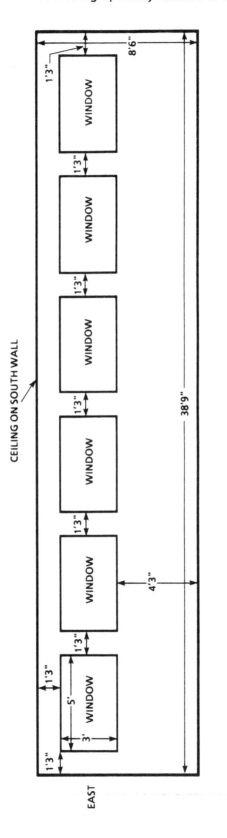

Teacher Information, Unit 9

ACTIVITY 3: Lever arm and force ratios

Procedure

a. and b. The output forces and input forces at the 40-cm fulcrum position are given in the data table below.

c. The output forces and input forces at the 60-cm fulcrum position are given in the data table below.

d. The output forces and input forces at the 80-cm fulcrum position are given in the data table below.

Fulcrum position	Input force	Output force
40 cm	1500 g	2250 g
40 cm	2500 g	3750 g
40 cm	3000 g	4500 g
60 cm	1500 g	1000 g
60 cm	2500 g	1667 g
60 cm	3000 g	2000 g
80 cm	1500 g	375 g
80 cm	2500 g	625 g
80 cm	3000 g	750 g

Calculations

a. *40-cm fulcrum position:*

$$\frac{\text{Input force}}{\text{Output force}} \stackrel{?}{=} \frac{1500\,g}{2250\,g} \stackrel{?}{=} \frac{2500\,g}{3750\,g} \stackrel{?}{=} \frac{3000\,g}{4500\,g} \stackrel{?}{=} \frac{2}{3}$$

60-cm fulcrum position:

$$\frac{\text{Input force}}{\text{Output force}} \stackrel{?}{=} \frac{1500\,g}{1000\,g} \stackrel{?}{=} \frac{2500\,g}{1667\,g} \stackrel{?}{=} \frac{3000\,g}{2000\,g} \stackrel{?}{=} \frac{3}{2}$$

80-cm fulcrum position:

$$\frac{\text{Input force}}{\text{Output force}} \stackrel{?}{=} \frac{1500\,g}{375\,g} \stackrel{?}{=} \frac{2500\,g}{625\,g} \stackrel{?}{=} \frac{3000\,g}{750\,g} \stackrel{?}{=} \frac{4}{1}$$

Yes, the ratios of input and output forces at the same fulcrum position are equal to each other. Yes, they form proportions.

b. No, the ratios of the input and output forces at different fulcrum positions are not equal to each other. No, they do not form proportions.

c. The following proportion holds true for all fulcrum positions.

$$\frac{\text{Input force}}{\text{Output force}} = \frac{\text{Output lever arm length}}{\text{Input lever arm length}} \quad \text{(an \textbf{inverse} proportion!)}$$

SOLUTIONS TO STUDENT EXERCISES

Note: The ratios shown as solutions below could have been constructed any of several ways. One possibility for each is given. The students may have a different construction, but should arrive at the same conclusion.

Exercise 1

a. The ratio of medical insurance deduction to gross pay is $51.01 : $427.50, or $\$51.01/\$427.50$.

b. $$\frac{\$51.01}{\$427.50} = \frac{y}{100}$$

   $y = (\$51.01 \times 100) \div \$427.50$
   $y = 11.93$ (rounded)

c. Approximately 11.93 percent (i.e., $11.93 out of each $100) is being deducted for medical insurance.

Exercise 2

The ratio of each component of the mix can be treated separately.

$$\frac{1}{1} = \frac{Cashews}{Almonds}$$

so,
Almonds = (Cashews × 1) ÷ 1
Almonds = Cashews

$$\frac{1}{3} = \frac{Cashews}{Peanuts}$$

so,
Peanuts = (Cashews × 3) ÷ 1
Peanuts = Cashews × 3

So, if there are 8 ounces of cashews, then

Almonds = Cashews = 8 ounces
Peanuts = Cashews × 3 = 24 ounces

Exercise 3

a. The ratio of drop of the light beam to the distance in front of the car permitted by the inspection program is 2" : 25', or 2 inches per 25 feet, or after converting to inches, 2 : 300, or 1 : 150.

b. $$\frac{1}{150} = \frac{28"}{y}$$

$y = (28" \times 150) \div 1$
$y = 4200"$, or 350' (that is, $4200" \times {}^{1'}/_{12"}$)

So, the car light beam will hit the ground at 350 feet in front of the car.

Exercise 4

a. 1. FICA withholding is normally found to be directly related to an individual's gross pay: higher gross pay means more FICA withholding.

    2. $$\frac{\$420.00}{\$30.03} \stackrel{?}{=} \frac{\$617.20}{\$44.13}$$

$420.00 \times 44.13 \stackrel{?}{=} 30.03 \times 617.20$
$18{,}535 = 18{,}535$

Yes, the two ratios are proportional.

b. 1. Air conditioner thermostat settings are normally found to be indirectly related to one's electric bill: a lower setting on the air conditioner results in a higher electric bill.

    2. $$\frac{68°}{\$103.12} \stackrel{?}{=} \frac{78°}{\$88.54}$$

$68 \times 88.54 \stackrel{?}{=} 78 \times 103.12$
$6020.72 \neq 8043.36$

No, the two ratios are *not* proportional.

One possible explanation for the ratios not being proportional might be that many other appliances are also used that affect the electricity usage, and hence the bill.

c. 1. Object height and length of object shadow are normally found to be directly related: the taller an object, the longer the shadow cast.

2. $$\frac{30'}{26'} \stackrel{?}{=} \frac{6'}{5.2'}$$

$30 \times 5.2 \stackrel{?}{=} 26 \times 6$
$156 = 156$

Yes, the two ratios are proportional.

d. 1. Size of drink is normally found to be inversely related to the cost—known as volume discount.

2. $$\frac{12 \text{ oz}}{\$0.50} \stackrel{?}{=} \frac{32 \text{ oz}}{\$0.89}$$

$12 \times 0.89 \stackrel{?}{=} \$0.50 \times 32$
$10.68 \neq 16.00$

No, the ratios are *not* proportional.

The sales price with volume discounts does not normally decrease with volume in a linear (or proportional) fashion, else eventually the drink would be free!

e. 1. Interest and average balance are normally directly related: a higher average balance normally means more interest will be earned.

2. $$\frac{\$4.80}{\$480.00} \stackrel{?}{=} \frac{\$11.70}{\$936.00}$$

$4.80 \times 936.00 \stackrel{?}{=} 11.70 \times 480.00$
$4492.80 \neq 5616.00$

No, the ratios are *not* proportional.

It may be that this institution awards customers with a higher interest rate for larger average balances.

Exercise 5

a. The dimensions in the picture should be directly related. While this is normally true for illustrations, it may not be true for "drawings" that are sometimes not drawn "to scale."

b. Illustration's dimensions:

$$\frac{\text{Shelf height in picture}}{\text{Hutch height in picture}}$$

Actual dimensions:

$$\frac{\text{Actual shelf height}}{\text{Actual hutch height}}$$

c. Substituting the known values, and solving for the unknown "Actual shelf height"...

$$\frac{\text{Shelf height in picture}}{\text{Hutch height in picture}} = \frac{\text{Actual shelf height}}{\text{Actual hutch height}}$$

$$\frac{11.4 \text{ cm}}{23.2 \text{ cm}} = \frac{y}{38.75''}$$

$y = (11.4 \text{ cm} \times 38.75'') \div 23.2 \text{ cm}$
$y = 19.04''$ (rounded)

So, the actual shelf height is about 19 inches, *not enough* room for your computer that is 21 $1/2''$ tall.

Exercise 6

a. Use the middle number, the phosphate content, to compare. The desired mix will have a nitrogen content that is 3 times the phosphate (3 : 1) and a potassium content that is 2 times the phosphate (2 : 1). If both of these ratios exist, then the mix will be in the desired ratio of 3 : 1 : 2.

   <u>Mix 1 (28-4-4)</u>

   Nitrogen: Nitrogen content $\stackrel{?}{=}$ 3 × Phosphate content
   $28 \stackrel{?}{=} 3 \times 4$
   $28 \neq 12$

   so this mix is not 3 : 1 : 2.

   <u>Mix 2 (21-0-0)</u>

   Nitrogen: Nitrogen content $\stackrel{?}{=}$ 3 × Phosphate content
   $21 \stackrel{?}{=} 3 \times 0$
   $21 \neq 0$

   so this mix is not 3 : 1 : 2.

Mix 3 (15-5-14)

    Nitrogen: Nitrogen content $\stackrel{?}{=}$ 3 × Phosphate content
                            15 $\stackrel{?}{=}$ 3 × 5
                            15 = 15

so check to see if potassium is in correct ratio.

    Potassium: Potassium content $\stackrel{?}{=}$ 2 × Phosphate content
                              14 $\stackrel{?}{=}$ 2 × 5
                              14 ≠ 10

so this mix is not 3 : 1 : 2.

Mix 4 (12-4-8)

    Nitrogen: Nitrogen content $\stackrel{?}{=}$ 3 × Phosphate content
                            12 $\stackrel{?}{=}$ 3 × 4
                            12 = 12

so check to see if potassium is in correct ratio.

    Potassium: Potassium content $\stackrel{?}{=}$ 2 × Phosphate content
                              8 $\stackrel{?}{=}$ 2 × 4
                              8 = 8

so *this mix is in the ratio of 3 : 1 : 2.*

Mix 5 (13-13-13)

    Nitrogen: Nitrogen content $\stackrel{?}{=}$ 3 × Phosphate content
                            13 $\stackrel{?}{=}$ 3 × 13
                            13 ≠ 39

so this mix is not 3 : 1 : 2.

Mix 6 (12-24-12)

    Nitrogen: Nitrogen content $\stackrel{?}{=}$ 3 × Phosphate content
                            12 $\stackrel{?}{=}$ 3 × 24
                            12 ≠ 72

so this mix is not 3 : 1 : 2.

So, only Mix 4 (12-4-8) has the three components in the proper ratio of 3 : 1 : 2.

Exercise 7

a.   Current ratio for this year = $47,206/$9498
      Current ratio for this year = 4.97 : 1 (rounded)

b. Current ratio for previous year = $38,314/$7908
   Current ratio for previous year = 4.84 : 1 (rounded)

c. The Browns' financial position has *strengthened* this year, since their current ratio has increased.

Exercise 8

The ratio of the pulley speeds should be indirectly proportional to the ratio of the pulley diameters.

$$\frac{\text{Speed of drive pulley}}{\text{Speed of blower pulley}} = \frac{\text{Diameter of blower pulley}}{\text{Diameter of drive pulley}}$$

$$\frac{1100 \text{ rpm}}{650 \text{ rpm}} = \frac{y}{10''}$$

y = (1100 rpm × 10") ÷ 650 rpm
y = 16.9" (rounded)

You should use a 16.9" pulley on the forage blower or, more realistically, a 17" pulley.

Exercise 9

a. The ratio of full-strength herbicide to water is 1 : 10.

b. The proportion of the desired ratio and the specified ratio...

$$\frac{\text{Specified parts of herbicide}}{\text{Specified parts of water}} = \frac{\text{Desired measure of herbicide}}{\text{Desired measure of water}}$$

$$\frac{1 \text{ part}}{10 \text{ parts}} = \frac{y}{14 \text{ oz}}$$

y = (1 part × 14 oz) ÷ 10 parts
y = 1.4 oz

So, you should add 1.4 oz of full-strength herbicide to the 14 oz of water.

c. The ratio of full-strength herbicide to water is 1 oz to 1 gallon. A better way to report this is without units. Converting 1 gallon to 128 ounces, yields a ratio of 1 oz herbicide to 128 oz water, or 1 : 128.

d. The strength of the solution for use in the hose-end sprayer is much greater than that used for direct application.

Exercise 10

a. and b.

Pounds of hay	Pounds of grain	Change in grain	Marginal rate of substitution
1000	1316		
1100	1259	57	0.57
1200	1208	51	0.51
1300	1162	46	0.46
1400	1120	42	0.42
1500	1081	39	0.39
1600	1046	35	0.31
1700	1014	32	0.32
1800	984	30	0.30
1900	957	27	0.27

c. The price ratio of hay cost to grain cost is $6¢/15¢$, or 0.40.

d. From the table above, the lowest cost combination is 1400 lb hay and 1120 lb of grain, which has a marginal rate of substitution of 0.42, just slightly above the price ratio of 0.40.

Exercise 11

a. The ratio of statue width to height is 4.5 : 17 or $4.5/17$.

b. Since a photograph is a proportional scale of the real image, the proportion of the actual widths to the desired widths is

$$\frac{\text{Actual height}}{\text{Actual width}} = \frac{\text{Photo height}}{\text{Photo width}}$$

$$\frac{17'}{4.5'} = \frac{y}{4"}$$

$$y = (17' \times 4") \div 4.5'$$
$$y = 15.1" \text{ (rounded)}$$

So the photo will require a little more than 15" in height, if it is to fill the 4" width.

Exercise 12

a. $$\text{Current ratio} = \frac{\text{Total current assets}}{\text{Total current liabilities}}$$

$$\text{Current ratio} = \frac{\$32{,}625}{\$19{,}480}$$

Current ratio = 1.67 (rounded)

So, your current ratio is about 1.67 to 1, or 1.67 : 1.

b. A bank would probably *not* want to loan you any more money, since your current ratio is less than 2 to 1.

Exercise 13

a. Average inventory = (Beginning inventory + Ending inventory) ÷ 2
Average inventory = ($28,437 + $33,010) ÷ 2
Average inventory = $30,723.50

b. $$\text{Inventory turnover} = \frac{\$263{,}841}{\$30{,}723.50}$$

Inventory turnover = 8.59 (rounded)

So, the inventory turnover is about 8.59 to 1, or 8.59 : 1.

c. The business is a *low inventory, fast turnover* business because the inventory turnover ratio is greater than 1 to 1.

Exercise 14

First total the investments in the business.

$22,000 + $48,000 = $70,000

Then determine the ratio of each man's investment to the total investment.

West's ratio = $22{,}000/70{,}000$

Brown's ratio = $48{,}000/70{,}000$

Then use these ratios to set up and solve the proportion that equates each ratio to the ratio of the share of the profits to the total profits.

$$\frac{22{,}000}{70{,}000} = \frac{\text{West's share}}{\$89{,}600}$$

West's share = $(22{,}000 \times \$89{,}600) \div 70{,}000$
West's share = $\$28{,}160$

$$\frac{48{,}000}{70{,}000} = \frac{\text{Brown's share}}{\$89{,}600}$$

Brown's share = $(48{,}000 \times \$89{,}600) \div 70{,}000$
Brown's share = $\$61{,}440$

Exercise 15

a. and b. The student's sketch should look similar to the one below. The ratio of the width to the height of the inner rectangle should be 4 : 3, and the entire rectangle 2 : 1, as stated in the problem.

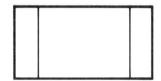

c. The students should measure the portion of the wide-screen movie width that is lost, and compare this to the total width of the wide-screen movie. For example, if the student's screen height is drawn as 3", then the standard aspect ratio of 4 : 3 would yield a screen width of 4" ($^4\!/_3 \times 3"$). For the 2 : 1 picture, the width would be 6" ($^2\!/_1 \times 3"$).

Picture lost = Full width of picture − Width of screen
Picture lost = 6" − 4"
Picture lost = 2"

Percent of picture lost = (Picture lost ÷ Picture width) × 100%
Percent of picture lost = (2" ÷ 6") × 100%
Percent of picture lost = 33.33% (rounded)

d. The full height of the TV screen would not be used. There would be blank space at the top and/or bottom of the screen. For example, if the picture width were constrained to the 4" sketch, then the height with the 2 : 1 ratio...

$$\frac{4"}{\text{Picture height}} = \frac{2}{1}$$

Picture height = (4" × 1) ÷ 2
Picture height = 2"

And so, with the 3" screen, there would be 1" blank space (3" – 2").

Exercise 16

a. The ratios, rounded to the nearest tenth...

$230 \text{ hr}/6 \text{ typists} = 38.3$ hr per typist

$253 \text{ hr}/7 \text{ typists} = 36.1$ hr per typist

$272 \text{ hr}/8 \text{ typists} = 34.0$ hr per typist

$285 \text{ hr}/9 \text{ typists} = 31.7$ hr per typist

$310 \text{ hr}/10 \text{ typists} = 31.0$ hr per typist

b. The trend shows that as the number of typists increases, the terminal use per typist decreases.

c. There is an indirect relationship between the number of typists and the time each typist spends at the terminals.

Exercise 17

a. The ratio of the width to the length is 841 : 1189, or $841/1189$, or 0.707 as a decimal value.

b. Based on the drawing in the text, cutting the A0 paper in half would yield a width of 594.5 mm ($1189 \text{ mm}/2$), and a length of 841 mm. The ratio of width to length is 594.5 : 841, or $594.5/841$, or 0.707 as a decimal value. This is the same ratio possessed by the A0 paper!

c. As indicated in the text of the exercise, each smaller size can be generated by folding the larger size in half. Thus the A0 paper can be folded in half, and then folded in half again to yield a card with A2 dimensions. The ratio of the card's width to height will still be 0.707, an interesting feature of the ISO paper sizes.

Exercise 18

The ratio of each model's inventory can be reduced by a factor of 4.

Selvo: $12/4 = 3$
Thistle: $28/4 = 7$
Cascade: $20/4 = 5$
Luxurie: $8/4 = 2$

Yes, the ratio of models on the lot is in agreement with the recommended ratio.

Exercise 19

a. Total of all ages is $52 + 48 + 33 = 133$

   Heir 1: $52 : 133$, or $52/133$
   Heir 2: $48 : 133$, or $48/133$
   Heir 3: $33 : 133$, or $33/133$

b. Heir 1:
   $$\frac{52}{133} = \frac{\text{Inheritance}}{\$107,065}$$
   Inheritance $= (52 \times \$107,065) \div 133$
   Inheritance $= \$41,860$

   Heir 2:
   $$\frac{48}{133} = \frac{\text{Inheritance}}{\$107,065}$$
   Inheritance $= (48 \times \$107,065) \div 133$
   Inheritance $= \$38,640$

   Heir 3:
   $$\frac{33}{133} = \frac{\text{Inheritance}}{\$107,065}$$
   Inheritance $= (33 \times \$107,065) \div 133$
   Inheritance $= \$26,565$

Solutions to Student Exercises

c. You can check your calculations by adding the inheritances to see if all the estate was dispersed:

Total inheritance = $\text{Heir}_1 + \text{Heir}_2 + \text{Heir}_3$
Total inheritance = $41,860 + $38,640 + $26,565
Total inheritance = $107,065

which agrees with the total given.

Exercise 20

a. 900 parts : 5 workers, simplifies to 180 parts : 1 worker, which is the 4-hour rate per worker: 180 parts per worker.

b. 2500 parts − 900 parts = 1600 parts

c. A proportion is set up, equating the ratio of number of workers to their output for the first four hours, to the ratio of the number employees needed to the output desired for the second four hours.

$$\frac{5 \text{ workers}}{900 \text{ parts}} = \frac{N \text{ workers}}{1600 \text{ parts}}$$

N workers = (5 workers × 1600 parts) ÷ 900 parts
N workers = 8.89 workers, or 9 workers (rounded up)

d. You will need to reassign 4 additional workers to the project.

Exercise 21

a. The ratio of salt to water in the solution is 1 : 25.

b. Mixing 1 part salt and 25 parts water would yield 26 parts total solution.

$$\frac{1 \text{ part salt}}{26 \text{ total parts}} = \frac{N \text{ parts salt}}{100 \text{ parts solution}}$$

N parts salt = (1 part salt × 100 parts solution) ÷ 26 total parts
N parts salt = 3.85 parts (rounded)

c. Since 3.85 parts per 100 are salt, the solution is 3.85 percent salt.

Exercise 22

a. The ratio of red blood cells to all other cells is 99.98 : 0.02, or $^{99.98}/_{0.02}$, or 4999.00 as a decimal value.

b. Set up a ratio of the observed cell counts, and compare to the desired ratio. The observed ratio is 4,600,000 : 6000, or $^{4,600,000}/_{6000}$, or 766.67 (rounded).

  The blood count is *not* close to being normal. The count of other cells is *higher* than it should be (over 6 times as high).

Exercise 23

a. The ratio of flu shots to patients in the first year is 622 : 4836, or $^{622}/_{4836}$.
  The ratio of flu shots to patients in last six months is 481 : 3418, or $^{481}/_{3418}$.

b. Evaluate the ratios as decimal fractions to simplify the fraction.
  First year = $^{622}/_{4836}$
  First year = 0.129 flu shots per patient seen

  First six months = $^{481}/_{3418}$
  First six months = 0.141 flu shots per patient seen

  Note that the question involves the number of flu shots per patient seen. This requires the proper choice of numerator (flu shots given) and denominator (patients seen) to easily be able to make the comparison. The decimal fraction for the second ratio is greater than the first ratio. Therefore, the first six months of this year the clinic is giving proportionally *more* shots than the first year it was open.

Exercise 24

a. Rounded to three decimal places...

  Daytime calls:
  $^{2.7\,mi}/_{5.5\,min}$ = 0.491 mi per min
  $^{6.2\,mi}/_{11.1\,min}$ = 0.559 mi per min
  $^{3.9\,mi}/_{8.8\,min}$ = 0.443 mi per min

  Nighttime calls:
  $^{4.0\,mi}/_{6.2\,min}$ = 0.645 mi per min
  $^{8.3\,mi}/_{10\,min}$ = 0.830 mi per min
  $^{2.1\,mi}/_{3.3\,min}$ = 0.636 mi per min

b. The data does show a faster response time (i.e., faster rate of travel) for nighttime calls than for daytime calls.

c. Average daytime rate = Sum of rates ÷ Number of values
Average daytime rate = (0.491 mi/min + 0.559 mi/min + 0.443 mi/min) ÷ 3
Average daytime rate = 0.498 mi per min (rounded)

Average nighttime rate = Sum of rates ÷ Number of values
Average nighttime rate = (0.645 mi/min + 0.830 mi/min + 0.636 mi/min) ÷ 3
Average nighttime rate = 0.704 mi per min (rounded)

The ratio of nighttime rate to daytime rate is 0.704 : 0.498, or $0.704/0.498$, or as a decimal fraction, 1.41 (rounded).

d. "When answering nighttime calls, the crew can drive to the scene about *1.4* (roughly one and one half) times faster than on daytime calls."

Exercise 25

a. The desired ratio of chemical to water is 0.9 : 1,000,000, or $0.9/1,000,000$.

b. The desired ratio of chemical to water should be equal to the ratio of chemical to water in the holding tank.

$$\frac{0.9 \text{ parts}}{1,000,000 \text{ parts}} = \frac{N \text{ gal chemical}}{700,000 \text{ gal tank}}$$

N gallons chemical = (0.9 parts × 700,000 gal) ÷ 1,000,000 parts
N gallons chemical = 0.63 gal

Exercise 26

The normal ratio is 5 gal : 75 guests, or $5 \text{ gal}/75 \text{ guests}$. Since you expect proportionally more punch to be needed for the larger crowd, you can equate this to the ratio for the larger crowd.

$$\frac{5 \text{ gal}}{75 \text{ guests}} = \frac{N \text{ gallons}}{300 \text{ guests}}$$

N gallons = (5 gal × 300 guests) ÷ 75 guests
N gallons = 20 gal

Exercise 27

a. The ratio of neck size to chest size for smallest shirt is 37 : 86, or $^{37}/_{86}$. The ratio of neck size to chest size for the largest shirt is 46 : 122, or $^{46}/_{122}$.

b. Check for equal ratios of the two extremes.

$$\frac{37}{86} \stackrel{?}{=} \frac{46}{122}$$
$$37 \times 122 \stackrel{?}{=} 86 \times 46$$
$$4514 \neq 3956$$

No, the ratio is *not the same* for all sizes.

Exercise 28

a. Calories from fat = Grams of fat × 9 calories per gram
Calories from fat = 6 g × 9 cal per g
Calories from fat = 54 cal

Ratio of fat-contributed calories to total calories is 54 : 130, or $^{54}/_{130}$.

b. Calories from fat = Grams of fat × 9 calories per gram
Calories from fat = 4 g × 9 cal per g
Calories from fat = 36 cal

Ratio of fat-contributed calories to total calories is 36 : 110, or $^{36}/_{110}$.

c. Cereal A = $^{54}/_{130}$ = 0.42

Cereal B = $^{36}/_{110}$ = 0.33

Cereal A (Part a) has proportionally more of the calories coming from the fat content of the cereal than Cereal B (Part b).

Exercise 29

a. Must first convert to the same units.

Ounces = Gallons × 128 ounces per gallon
Ounces = 1 gal × 128 oz per gal
Ounces = 128 oz

Ratio of concentrate to water is 5 : 128, or $^{5}/_{128}$.

b. Set up and solve the proportion.

$$\frac{5}{128} = \frac{\text{Amount of concentrate}}{32 \text{ oz}}$$

Amount of concentrate = (5 × 32 oz) ÷ 128
Amount of concentrate = 1.25 oz

Exercise 30

a. To compute a child's share...

$$\frac{\text{Child's age}}{\text{Total of ages}} = \frac{\text{Share of deposit}}{\text{Total deposit}}$$

The first year, the total of their ages is 9 years. For the oldest child,

$$\frac{7 \text{ yr}}{9 \text{ yr}} = \frac{\text{Share of deposit}}{\$100}$$

Share of deposit = (7 yr × $100) ÷ 9 yr
Share of deposit = $77.78 (rounded)

For the youngest child,

$$\frac{2 \text{ yr}}{9 \text{ yr}} = \frac{\text{Share of deposit}}{\$100}$$

Share of deposit = (2 yr × $100) ÷ 9 yr
Share of deposit = $22.22 (rounded)

b. For the 12th year of the plan, the oldest child will be 18 years old and the youngest will be 13 years old. The total of their ages is 31 years.

For the oldest child,

$$\frac{18 \text{ yr}}{31 \text{ yr}} = \frac{\text{Share of deposit}}{\$100}$$

Share of deposit = (18 yr × $100) ÷ 31 yr
Share of deposit = $58.06 (rounded)

For the youngest child,

$$\frac{13 \text{ yr}}{31 \text{ yr}} = \frac{\text{Share of deposit}}{\$100}$$

Share of deposit = (13 yr × $100) ÷ 31 yr
Share of deposit = $41.94 (rounded)

c. Continuing to contribute proportionally to each child's account by age will mean that the oldest child will always receive more money. The difference between the amount of the oldest and youngest children's payments will decrease with time. If you stop payments to the older child's account at age 19, and then contribute the total amount to the youngest child's account, the younger child will end up with slightly more money. This is probably still fair since inflation will likely increase college costs.

For students who need extra challenge, you may suggest that they create a table of such a payment schedule up to the 19th birthday of the youngest child. The students could show the total of the monthly payments for each year and the cumulative balances of each account. You may also suggest that they apply a 10% lump-interest payment on the end of the year balance. Would they like to have such an account established for themselves?

Exercise 31

Since the pulley diameters and speeds are inversely proportional

$$\frac{\text{Diameter A}}{\text{Diameter B}} = \frac{\text{Speed B}}{\text{Speed A}}$$

$$\frac{2"}{5"} = \frac{\text{Speed B}}{1075 \text{ rpm}}$$

Speed B = (2" × 1075 rpm) ÷ 5"
Speed B = 430 rpm

The fan blade is mounted on the same shaft as the 5" pulley so it will make the same number of revolutions per minute—430 rpm.

Exercise 32

Since the gear ratios are directly proportional to the speeds

$$\frac{3.4}{1} = \frac{\text{Crankshaft speed}}{\text{Drive axle speed}}$$

Drive axle speed = (1 × 1500 rpm) ÷ 3.4
Drive axle speed = 441 rpm (rounded)

Solutions to Student Exercises

Exercise 33

a. The oils are to be mixed in the ratio of 8 to 15.

$$\frac{8}{15} = \frac{16 \text{ oz of Code 171}}{\text{Amount of Code 209}}$$

Amount of Code 209 = (15 × 16 oz) ÷ 8
Amount of Code 209 = 30 oz

b. Total batch = 16 oz + 30 oz
Total batch = 46 oz

c. Fraction of Code 171 = $16 \text{ oz}/46 \text{ oz}$
Fraction of Code 171 = 0.348 (rounded)
Fraction of Code 171 = 34.8% (as a percentage)

Fraction of Code 209 = $30 \text{ oz}/46 \text{ oz}$
Fraction of Code 209 = 0.652 (rounded)
Fraction of Code 209 = 65.2% (as a percentage)

Exercise 34

a. The relation between the number of turns and the voltage for each coil is

$$\frac{10{,}000 \text{ turns}}{120 \text{ V}} = \frac{1000 \text{ turns}}{12 \text{ V}}$$

b. Since the ratios are directly paired (number of turns and voltages on both sides of the equality), the primary and secondary voltages are directly proportional to the number of turns of wire.

c. If 120 V are applied to the 1000 turns,

$$\frac{1000 \text{ turns}}{120 \text{ V}} = \frac{10{,}000 \text{ turns}}{\text{Output voltage}}$$

Output voltage = (120 V × 10,000 turns) ÷ 1000 turns
Output voltage = 1200 V

Exercise 35

a. Using the prescribed ratio for cement to sand of 1 : 3,

$$\frac{1 \text{ part cement}}{3 \text{ parts sand}} = \frac{0.05 \text{ m}^3}{\text{Volume of sand}}$$

Volume of sand = (3 parts × 0.05 m³) ÷ 1 part
Volume of sand = 0.15 m³

b. Amount of one batch = 0.05 m³ + 0.15 m³
Amount of one batch = 0.2 m³

Mixer will hold = 0.65 m³ ÷ 0.2 m³ per batch
Mixer will hold = 3.25 batches, or 3 batches (rounded to nearest batch)

Exercise 36

a. The ratio of drawing dimension to the actual dimension is 1/4 inch = 1 foot or, changing to the same units, the ratio can be expressed as 0.25 : 12, or 1 : 48. Thus the drawing is 1/48 scale.

b. To represent a 24' wall with this scale,

$$\frac{1}{48} = \frac{\text{Drawing dimension}}{24' \text{ wall}}$$

Drawing dimension = (1 × 24' wall) ÷ 48
Drawing dimension = 0.5', or 6" line to depict the wall.

c. The actual dimensions of a duct shown as 9 1/2" on such a drawing:

$$\frac{1}{48} = \frac{9\frac{1}{2}"}{\text{Actual dimension}}$$

Actual dimension = (9 1/2" × 48) ÷ 1
Actual dimension = 456", or 38' for the actual length of the duct.

Exercise 37

a. First, the number of revolutions of the drum (i.e., the large gear) must be determined.

Circumference of drum = πd
Circumference of drum = 3.14 × 1.5"
Circumference of drum = 4.71"

So, since each revolution will wrap about 4.71",

Solutions to Student Exercises

Number of revolutions = Required distance ÷ Circumference per revolution
Number of revolutions = 12" ÷ 4.71" per revolution
Number of revolutions = 2.55 revolutions

Then, using the given ratio of the gear

$$\frac{1 \text{ rev}}{41 \text{ turns}} = \frac{2.55 \text{ rev}}{\text{Number of crank turns}}$$

Number of crank turns = (2.55 rev × 41 turns) ÷ 1 rev
Number of crank turns = 105 turns (rounded)

b. The rate of turning (40 turns : 1 min) can be handled as a ratio.

$$\frac{40 \text{ turns}}{1 \text{ min}} = \frac{105 \text{ turns}}{\text{Time needed}}$$

Time needed = (105 turns × 1 min) ÷ 40 turns
Time needed = 2.6 min (rounded)

Exercise 38

a. As stated, the pressure and area of the pistons are inversely proportional. This can be stated as shown below.

$$\frac{P_i}{P_o} = \frac{A_o}{A_i}$$

So for the pistons given in this exercise,

$$\frac{35 \text{ psi}}{P_o} = \frac{4 \text{ in}^2}{25 \text{ in}^2}$$

P_o (output pressure) = (35 psi × 25 in²) ÷ 4 in²
P_o (output pressure) = 218.75 psi

b. Suppose that the input pressure (P_i) was doubled (increased to 70 psi).

$$\frac{70 \text{ psi}}{P_o} = \frac{4 \text{ in}^2}{25 \text{ in}^2}$$

P_o (output pressure) = (70 psi × 25 in²) ÷ 4 in²
P_o (output pressure) = 437.5 psi

This is 2 times the previous output pressure. So if the input pressure is doubled, the output pressure is doubled. The output pressure is directly proportional to the input pressure.

c. Suppose the area of the output piston (A_o) is doubled to 8 in².

$$\frac{35 \text{ psi}}{P_o} = \frac{8 \text{ in}^2}{25 \text{ in}^2}$$

P_o (output pressure) = (35 psi × 25 in²) ÷ 8 in²
P_o (output pressure) = 109.375 psi

This is ½ of the previous output pressure. So if the area of the output piston is doubled, the output pressure is halved. The output pressure is inversely proportional to the area of the output piston.

Exercise 39

a. The rising grade is a ratio of rise to run of 1 : 42, or ¹⁄₄₂, or 0.0238. The descending grade is a ratio of 1 : 78, or ¹⁄₇₈, or 0.0128. The 3-mile rising grade has the greater rise to run ratio (0.0238 > 0.0128).

b. A proportion can be set up with the ratio of the rising grade. Anticipating an answer in feet, use the rising distance as 15,840 feet (3 mi × 5280 ft per mi).

$$\frac{1}{42} = \frac{\text{Altitude gain}}{15,840' \text{ distance}}$$

Altitude gain = (1 × 15,840') ÷ 42
Altitude gain = 377.14 ft (rounded)

c. Similarly, a proportion can be set up with the descending grade, using a distance of 26,400 feet (5 mi × 5280 ft per mi).

$$\frac{1}{78} = \frac{\text{Altitude gain}}{26,400' \text{ distance}}$$

Altitude gain = (1 × 26,400') ÷ 78
Altitude gain = 338.46 ft (rounded)

No, the train does not lose all of the altitude it gained on the rising grade. It is still roughly 39 feet above the starting altitude.

Exercise 40

a. The compression ratio is 8 to 1.

b. The domed top is included in both the beginning and ending volumes. From the drawing, the diameter of the hemisphere is 6.250 inches.

Stroke volume $= \pi r^2 h$, $r = {}^d/_2$
Stroke volume $= 3.14 \times (6.250 \text{ in} \div 2)^2 \times 8.000''$
Stroke volume $= 245.31 \text{ in}^3$ (rounded)

Beginning volume $=$ Stroke volume $+$ Top combustion volume
Beginning volume $= 245.31 \text{ in}^3 + 15.970 \text{ in}^3$
Beginning volume $= 261.28 \text{ in}^3$

Ending volume $=$ Top combustion volume
Ending volume $= 15.970 \text{ in}^3$

$$\text{Compression ratio} = \frac{\text{Beginning volume}}{\text{Ending volume}}$$

$$\text{Compression ratio} = \frac{261.28 \text{ in}^3}{15.970 \text{ in}^3}$$

$$\text{Compression ratio} = \frac{16.36}{1} \text{ (rounded)}$$

Thus the compression ratio is 16.36 to 1.

STUDENT HANDOUTS

The following pages are copies of student resources that were introduced in the student text for Unit 3, "Measuring in English and Metric Units." These resources include "Tables of Conversion Factors" and "How to Use the Accu-Line™ Drawing Aid."

If your students did not complete Unit 3, or if the students did not keep copies of that text, you may copy and distribute these aids to your students as needed—to complete the laboratory activities and the student exercises included in this unit. While parts of these resources may not apply to this unit, all or part of these resources will be used in Units 4-22 of this series. Therefore, the resources are supplied in their entirety so that you may use them as you need them.

TABLES OF CONVERSION FACTORS

To convert from meters to inches, for example, find the row labeled "1 meter" and the column labeled "in." The conversion factor is 39.37. Thus, 1 meter = 39.37 in.

LENGTH

	cm	m	km	in.	ft	yd	mi
1 centimeter	1	0.01	10^{-5}	0.3937	3.281×10^{-2}	1.094×10^{-2}	6.214×10^{-6}
1 meter	100	1	10^{-3}	39.37	3.281	1.094	6.214×10^{-4}
1 kilometer	10^5	1000	1	3.937×10^4	3281	1094	0.6214
1 inch	2.54	0.0254	2.54×10^{-5}	1	0.0833	0.0278	1.578×10^{-5}
1 foot	30.48	0.3048	3.048×10^{-4}	12	1	0.3333	1.894×10^{-4}
1 yard	91.44	0.9144	9.144×10^{-4}	36	3	1	5.682×10^{-4}
1 mile	1.6093×10^5	1609.3	1.6093	6.336×10^4	5280	1760	1

AREA

	cm^2	m^2	$in.^2$	ft^2	A	mi^2
1 square centimeter	1	10^{-4}	0.1550	1.076×10^{-3}	2.471×10^{-8}	3.861×10^{-11}
1 square meter	10^4	1	1550	10.76	2.471×10^{-4}	3.861×10^{-7}
1 square inch	6.452	6.452×10^{-4}	1	6.944×10^{-3}	1.594×10^{-7}	2.491×10^{-10}
1 square foot	929.0	0.09290	144	1	2.296×10^{-5}	3.587×10^{-8}
1 acre	4.047×10^7	4047	6.273×10^6	43,560	1	1.563×10^{-3}
1 square mile	2.590×10^{10}	2.590×10^6	4.007×10^9	2.788×10^7	640	1

NOTE: The use of "scientific notation" will be discussed in Unit 12 of Applied Mathematics. Briefly, the number 3.281×10^{-2} is equivalent to 0.03281, and the number 3.937×10^4 is equivalent to 39,370. If you have difficulty understanding this, please consult your teacher.

VOLUME (CAPACITY)

	cm³	m³	in.³	ft³	ℓ	oz	gal
1 cubic centimeter	1	10^{-6}	0.06102	3.531×10^{-5}	1.000×10^{-3}	0.03381	2.642×10^{-4}
1 cubic meter	10^6	1	6.102×10^4	35.31	1000.	3.381×10^4	264.2
1 cubic inch	16.39	1.639×10^{-5}	1	5.787×10^{-4}	0.01639	0.5541	4.329×10^{-3}
1 cubic foot	2.832×10^4	0.02832	1728	1	28.32	957.5	7.480
1 liter	1000.	1.000×10^{-3}	61.03	0.03532	1	33.81	0.2642
1 ounce	29.57	2.957×10^{-5}	1.805	1.044×10^{-3}	0.02957	1	7.813×10^{-3}
1 gallon	3785	3.785×10^{-3}	231	0.1337	3.785	128	1

1 gallon = 4 quarts (qt) = 8 pints (pt) = 16 cups (c)
1 cup (c) = 8 ounces (oz) = 16 tablespoons (tbsp) = 48 teaspoons (tsp)

MASS / WEIGHT

	g	kg	oz	lb	ton*
1 gram	1	10^{-3}	0.03527	2.205×10^{-3}	1.102×10^{-6}
1 kilogram	10^3	1	35.27	2.205	1.102×10^{-3}
1 ounce	28.35	0.02835	1	0.0625	3.125×10^{-5}
1 pound	453.6	0.4536	16	1	0.0005
1 ton*	9.072×10^5	907.2	3.2×10^4	2000	1

*short ton

NOTE: The use of "scientific notation" will be discussed in Unit 12 of Applied Mathematics. Briefly, the number 3.281×10^{-2} is equivalent to 0.03281, and the number 3.937×10^4 is equivalent to 39,370. If you have difficulty understanding this, please consult your teacher.

Student Handouts

ANGLE

	′	°	rad	rev
1 minute	1	0.01667	2.909×10^{-4}	4.630×10^{-5}
1 degree	60	1	0.01745	2.778×10^{-3}
1 radian	3438	57.30	1	0.1592
1 revolution	2.16×10^4	360	6.283	1

TIME

	sec	min	h	d*	y*
1 second	1	0.01667	2.778×10^{-4}	1.157×10^{-5}	3.169×10^{-8}
1 minute	60	1	0.01667	6.944×10^{-4}	1.901×10^{-6}
1 hour	3600	60	1	0.04167	1.141×10^{-4}
1 day*	8.640×10^4	1440	24	1	2.738×10^{-3}
1 year*	3.156×10^7	5.259×10^5	8766	365.3	1

*sidereal

NOTE: The use of "scientific notation" will be discussed in Unit 12 of Applied Mathematics. Briefly, the number 3.281×10^{-2} is equivalent to 0.03281, and the number 3.937×10^4 is equivalent to 39,370. If you have difficulty understanding this, please consult your teacher.

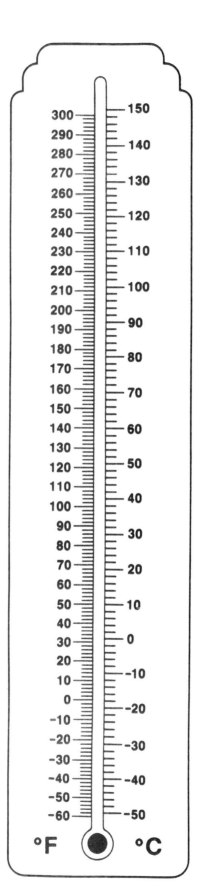

Dual-scale Thermometer

How to Use the Accu-Line™ Drawing Aid

The Accu-Line™ is a drawing aid. It is useful in making sketches with perpendicular and parallel lines and drawing angles with a specified measure.

The Accu-Line™ guides a ball-point pen or a mechanical pencil in narrow grooves along straight lines. With the Accu-Line™, you can draw parallel lines as close as 0.5 mm.

To draw a vertical line, place the point of the pen or pencil at the starting point for the line. Tilt the top of the pen or pencil toward the bottom of the pad and pull the pencil in the direction you are drawing, pressing on the paper with a normal writing pressure. See illustration **A** below.

To draw a horizontal line, place the point of the pen or pencil at the starting point for the line. Tilt the top of the pen or pencil to the right if you are right handed and to the left if you are left handed. **Pull** the pen or pencil in the direction you are drawing, pressing on the paper with a normal writing pressure. See illustration **B** below.

Always tilt the pen or pencil in the direction you are drawing. Always pull the pen or pencil in the direction you are drawing. Never push the pen or pencil.

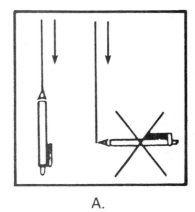
A.

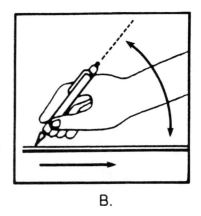
B.

Practice using the Accu-Line™ by sketching the floor plan of your home. Do not worry about the scale. You only need to show the layout of the rooms.

The Accu-Line™ is also useful for drawing angles. There is a protractor printed on the Accu-Line™. To draw angles, the paper you are drawing on must be removed from the pad. First, align the left edge of the paper with the vertical scale on the left side of the Accu-Line™. See illustration C. Draw a vertical reference line near this reference edge of the paper. Next, without moving the paper, draw a horizontal line on the paper. This line will be one of the sides of the angle you want to draw. Now, rotate the paper and align the reference edge of the paper on the specified angle measure of the protractor (for example, on the 30° mark for a 30° angle). See illustration D. Hold the paper firmly in position and draw a vertical line on the paper that crosses the horizontal line. The angle between the vertical line just drawn and the horizontal line is the desired angle.

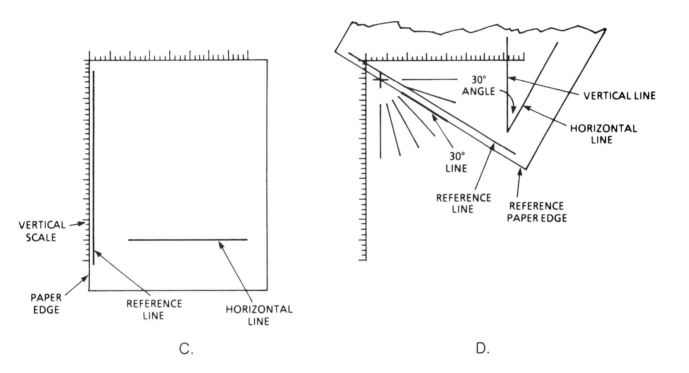

C.

D.

Practice drawing the following angles with the Accu-Line™.

Draw a 30° angle.

Draw a 45° angle.

Draw a 75° angle.

Practice using the Accu-Line™ to draw lines and angles of any shape.

SKILL DRILL
UNIT 9

1. Identify the expressions below that are ratios.
   a. 2 to 1
   b. 55 miles per hour
   c. 48 centimeters
   d. 1 inch : 1 foot
   e. $^{21}/_{14}$
   f. 32°F
   g. 22¢ per mile
   h. πr^2
   i. 24 : 4 : 8
   j. 2 × 4

For Questions 2 through 4, refer to the following statement: "Out of the sample of seeds you planted, 13 sprouted and 7 didn't."

2. What is the ratio of seeds that sprouted to seeds that didn't sprout?

3. What is the ratio of seeds that sprouted to the seeds planted?

4. If you planted a sack of 300 of these seeds under the same conditions, how many would you expect to sprout?

For Questions 5 through 7, refer to the following statement: "To obtain the *Sky Blue* paint color, add $^1/_2$ oz of pigment #17 to one gallon of white base paint."

5. What is the specified ratio of pigment to white base?

6. How much pigment would be used to color 10 gallons of white base to Sky Blue?

7. If you have 35 oz of pigment #17 on hand, how many gallons of Sky Blue paint can you make?

Convert the ratios given below to their simplest form. Convert to the same units, where possible.

8. 8 to 4

9. 100 : 75

10. 90 miles in 2 hours

11. 35 pounds of beans from 7 plants

12. 3 fluid oz concentrate in 15 gal of water

13. 6 inches on drawing is labeled as 24 feet

Determine whether the pairs of ratios below are equal or not.

14. $\dfrac{25}{10} \stackrel{?}{=} \dfrac{35}{14}$

15. $\dfrac{12}{21} \stackrel{?}{=} \dfrac{56}{98}$

16. $\dfrac{17}{48} \stackrel{?}{=} \dfrac{22}{64}$

17. $\dfrac{2.54}{12} \stackrel{?}{=} \dfrac{39.37}{186}$

18. $\dfrac{144}{3.14} \stackrel{?}{=} \dfrac{27}{0.875}$

19. $\dfrac{0.125}{4.5} \stackrel{?}{=} \dfrac{15}{540}$

Examine the pair of figures in each problem below. Determine whether the figures are "similar" or not.

20.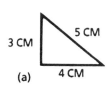

21.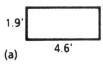

22.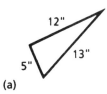

Skill Drill

T-51

Solve the proportions below to find the unknown term, y.

23. $\dfrac{24}{64} = \dfrac{y}{8}$

24. $\dfrac{4}{128} = \dfrac{64}{y}$

25. $\dfrac{2.54 \text{ cm}}{1 \text{ in.}} = \dfrac{114.3 \text{ cm}}{y}$

26. $\dfrac{100 \text{ mg}}{70 \text{ lb}} = \dfrac{y}{175 \text{ lb}}$

For the problems described below, identify whether the relationship is direct or indirect. Set up a proportion describing the relationship and solve for the unknown value.

27. While driving on a trip you are able to drive a total of 400 miles in ten hours. At this same driving rate, how many hours will it take you to drive 1750 miles?

28. At a certain time of day you cast a shadow that is 8 feet long, while the nearby shadow of a pine tree is 45 feet. You know your height is 5 feet. About how tall is the pine tree?

29. A trip on the interstate highway takes 2 hours when you travel 55 miles per hour. Your state has recently increased the speed limit to 65 miles per hour. How long do you estimate the same trip will take if you drive at 65 miles per hour?

30. A drive gear with 24 teeth meshes with a pinion gear with 60 teeth. If the drive gear is turning at a speed of 120 revolutions per minute, how fast is the pinion gear turning?

―――――――――――――ANSWER KEY―――――――――――――

1. a, b, d, e, g, i
2. 13 : 7
3. 13 : 20
4. 195 seeds
5. ½ oz to 1 gal
6. 5 oz
7. 70 gal
8. 2 to 1
9. 4 : 3
10. 45 miles per hour

11. 5 pounds per plant
12. 1 oz per 5 gal, or 1 : 640
13. 1" = 4', or 1 : 48
14. equal
15. equal
16. not equal
17. equal
18. not equal
19. equal
20. similar

21. not similar
22. similar
23. 3
24. 2048
25. 45 in.
26. 250 mg

27. A direct relationship

$$\frac{10 \text{ hr}}{400 \text{ mi}} = \frac{y}{1750 \text{ mi}}$$

y = 43.75 hr

28. A direct relationship

$$\frac{5'}{8'} = \frac{y}{45'}$$

y = 28.125'

29. An indirect relationship

$$\frac{55 \text{ mph}}{65 \text{ mph}} = \frac{y}{2 \text{ hr}}$$

y = 1.69 hr (rounded)

30. An indirect relationship

$$\frac{24 \text{ teeth}}{60 \text{ teeth}} = \frac{y}{120 \text{ rpm}}$$

y = 48 rpm

Skill Drill

PROBLEM BANK FOR END-OF-UNIT TEST

UNIT 9, "USING RATIOS AND PROPORTIONS"

1. Write the ratio of 3 inches to 1 1/4 feet in four different ways.

For Questions 2 through 4 refer to the following statistic. "The Blues won five of the first six games they played."

2. What was their win/loss ratio?

3. What was their ratio of wins to games played?

4. What was their ratio of losses to wins?

Read and solve the problems given in Questions 5 through 15.

5. If 4 inches on a blueprint represents 16 feet, how many feet does a line that is 12 inches long on the blueprint represent?

6. If twelve pairs of pliers cost $75, how much will 30 pairs cost?

7. If a truck travels 160 miles on 22 gallons of gasoline, how many gallons will it take to travel 400 miles?

8. What is the gear ratio (drive gear to pinion gear) if the drive gear has 24 teeth and the pinion gear has 60 teeth?

9. An oil burner uses 1200 gallons of oil in 2 months. Predict how much oil it will burn in 5 months.

10. One type of solder contains 40 percent tin and 60 percent lead. What is the ratio of lead to tin?

11. If 5 tomato plants yield 35 pounds of tomatoes, how many plants will be needed to yield 75 pounds?

12. In a gear train, the drive gear has 45 teeth and the pinion gear has 55 teeth. If the drive gear turns at an angular speed of 99 rpm, what is the angular speed of the pinion gear?

13. A pump moves 60 gallons of water in 4 hours. How much water will it move in 7 hours?

14. A recipe for a cheese and noodle casserole calls for 4 cups of noodles and 4 ounces of cheese. If you want to use 10 ounces of cheese, how many cups of noodles should you use for the casserole?

Problem Bank, Unit 9

15. A machine can make 17 parts in 5 minutes. How long will it take the machine to make 30 parts?

For Problems 16 through 20, compare the ratios given and classify them as equal or not equal.

16. $10:35$ and $2:7$
17. $14:28$ and $4:8$
18. $3/5$ and $12/20$
19. *4 inches to 12 inches* and *2 pounds to 6 pounds*
20. *4 inches to 6 pounds* and *2 inches to 12 pounds*
21. If a gear drive has a ratio of 4 to 1 (drive gear teeth to pinion gear teeth), how many times will the pinion gear turn if the drive gear turns 80 times?
22. An insecticide is prepared from a concentrate by mixing 1 part concentrate with 100 parts water. How much water (in gallons) should be mixed with 32 ounces of concentrate? (1 gallon = 128 ounces)

For Questions 23 through 26, identify the two quantities as directly or indirectly related.

23. Unit cost and total cost
24. Fuel burned and savings on fuel bill
25. Cups of flour used and cookies made
26. Speed of car and time needed to cover a distance
27. At 55 words per minute, how long will it take to type a 220-word letter?
28. If a pump circulates water from a 459,000-gallon swimming pool through filters at a rate of 850 gallons per minute, how many hours are required to filter all the water in the pool?
29. A freight elevator has a weight limit of 4500 pounds. How many drums of solvent weighing 435 pounds per drum can be carried by the elevator?
30. Water has a density (ratio of weight to volume) of 8 pounds per gallon. What is the volume of 440 pounds of water?

ANSWERS FOR PROBLEM BANK FOR END-OF-UNIT TEST

UNIT 9, "USING RATIOS AND PROPORTIONS"

1. 3 : 15; 3 to 15; $3/15$;
   3 inches per 15 inches
   or (1 : 5; 1 to 5; $1/5$;
   1 inch per 5 inches)
2. 5 : 1
3. 5 : 6
4. 1 : 5
5. 48 feet
6. $187.50
7. 55 gallons
8. $24/60$, $4/10$, $2/5$, or 0.4
9. 3000 gallons
10. 60 to 40 or (3 to 2)
11. 11 plants
12. 81 rpm
13. 105 gallons
14. 10 cups
15. 8.8 minutes
16. equal
17. equal
18. equal
19. equal
20. not equal
21. 320
22. 25 gallons
23. directly related
24. indirectly related
25. directly related
26. indirectly related
27. 4 minutes
28. 9.0 hours
29. 10 drums
30. 55 gallons

PART 2: STUDENT TEXT

Unit 9

Using Ratios and Proportions

Developed by the Center for Occupational Research and Development and sponsored by a consortium of State Vocational Education Agencies with the cooperation and support of mathematics educators.

©Copyright 1988 by the Center for Occupational Research and Development
Waco, Texas

Neither CORD nor any member of the consortium assumes any liabilities with respect to the use of, or for damages resulting from the use of, any information, apparatus, method or process described in these materials.

Vocational occupations icon design courtesy of the Vocational Education Services Project directed by Dale Law and John Smith of the University of Illinois.

Published and distributed by: CORD Communications
324 Kelly Drive, Waco, Texas 76710
817-776-1822 Fax 817-776-3906

Library of Congress Catalog Card Number: 87-072534
ISBN 1-55502-296-0 (Applied Mathematics)
ISBN 1-55502-309-6 (Unit 9, Using Ratios and Proportions)

PREFACE

Applied Mathematics contains video programs, laboratory activities and problem-solving exercises. Each part has been chosen to help you understand the mathematics you need to work and live in a technical world. Most importantly, each part has been designed to make mathematics more useful and meaningful for you—and reduce some of the "math anxiety" we all feel at one time or another.

Applied Mathematics contains some of the important mathematics you'll need to be a productive member of today's workforce. The ideas you'll learn will help you understand:
- numbers, decimals, fractions and percents;
- shapes and sizes;
- how to handle equations and formulas;
- how to work with angles and triangles;
- how to estimate answers and solve problems; and
- how to describe the behavior of large populations of things.

Each of these skills will help you do your job better—and help you advance to the next job when the time comes.

Each unit of **Applied Mathematics** begins with a video program. The video tells you about the mathematics skills you'll be studying and introduces you to real people who use these skills in their everyday lives. Following this, you'll concentrate on learning the mathematics skills.

Next, you'll have a chance to practice what you've learned. You'll do this in laboratory activities that involve measurement and problem solving. You'll apply the math skills to practical problems. These problems are the kind that people have to solve every day—in restaurants, on the farm, in a factory, in a business office, in a hospital, in a laboratory, or at home.

You'll have many opportunities to practice what you've learned, so don't be discouraged if it all doesn't seem too clear the first time around. You'll learn how to use mathematics, little by little, problem by problem. And in the process, you may find that using mathematics can be fun.

The CORD Project Staff

Table of Contents

LEARNING THE SKILLS

Introduction	3
Identifying Ratios	4
Reading and Interpreting Ratios	8
Comparing Ratios	10
Identifying and Writing Proportions	14
Solving Proportions	16
Summary	27

PRACTICING THE SKILLS

Laboratory Activities

Activity 1: Ratios and proportions in similar triangles	29
Activity 2: Using ratios to make scale drawings	31
Activity 3: Lever arm and force ratios	32

Student Exercises

General	34
Agriculture and Agribusiness	36
Business and Marketing	38
Health Occupations	43
Home Economics	44
Industrial Technology	46

REFERENCE MATERIALS

Glossary	51

Using Ratios and Proportions

How to use ratios and proportions to solve problems

Prerequisites	This unit builds on the skills taught in: Unit 1: *Learning Problem-solving Techniques* Unit 2: *Estimating Answers* Unit 3: *Measuring in English and Metric Units* Unit 7: *Working with Shapes in Two Dimensions*
To Master This Unit	Read the text and answer all the questions. Complete the assigned problems and activities. Work the problems on the unit test at a satisfactory level.
Unit Objectives	Working through this unit helps you learn how to: 1. Read and interpret ratios. 2. Compare ratios. 3. Recognize and write proportions from given information. 4. Distinguish between direct and indirect relationships. 5. Solve proportions in practical, work-related problems.

- Successful completion of this unit—and others in **Applied Mathematics**—requires that students use a scientific calculator to make calculations. See the annotated teacher notes for student page 5—in the **Teacher's Guide, Unit A, "Getting to Know Your Calculator"**—for comments on the type and cost of scientific calculators we recommend.
- In this unit we continue to use a versatile, durable and simple drawing tool that enables students to draw parallel lines, perpendicular lines and all angles quickly, and with great precision. It is called Accu-Line™. Like the scientific calculator, it will be used throughout the course. In this unit, they will use Accu-Line™ to help them complete Laboratory Activity 2. (An explanation of how to use Accu-Line™ is provided in the **Resources** section of this guide.)
- It's our opinion that a score of 75-85% on the Unit Test would indicate a reasonable mastery of the unit materials presented herein. However, the decision of what constitutes a "satisfactory level of performance" on the unit test has been left up to you.

Using Ratios
and
Proportions
Page 2

Learning Path

1. Read the "Introduction" section of this unit.

2. Watch the video and take part in the class discussions.

3. As you read the text, think about what the words say. Read the explanations and examples carefully. Use the Study Activities to help you practice as you read.

4. Do the assigned mathematics lab activity.

5. Complete the assigned exercises.

6. Measure your progress by taking the unit test.

Some Signals to Help You Learn

The following signals help you know what to do as you read the text:

 Think this through. Spend a little extra time on this idea.

 Write your answer on your paper.

 Carry out the calculations.

 Learn this key rule or definition.

 Estimate and ask yourself if this answer makes sense.

 Compare your answer to the given one and make any needed changes.

Applied Mathematics

- Students will probably skip over this page if they have already completed several units of **Applied Mathematics**. That's to be expected, because the material on this page is almost always the same. Periodically, however, you might review the meaning of the **icons** and point out that the **Learning Path** summarizes the main learning activities for the unit.

- Some of you may question the need for repeating this page, unit after unit. That's a reasonable question. It is necessary, though, despite its obvious redundancy, because the course in **Applied Mathematics** is to be used not only as a stand-alone course—with all units covered—but also as a set of learning materials with units used somewhat arbitrarily, out of sequence, and only as needed for special learning situations. As a result, we find it necessary to repeat this page, for the teachers and learners who may be using this unit in **Applied Mathematics** for the first time, without benefit of explanations given in previous units.

LEARNING THE SKILLS

INTRODUCTION

The mathematics that helps you compare two quantities is very useful. When you say that the score for yesterday's baseball game was "3 to 2," you are comparing one team's points to the other. The comparison "3 to 2" is called a **ratio**.

When you read distances on a map, or measurements on a blueprint, you are using ratios. Builders speak of the *pitch of a roof* and road crews talk about the *grade of a hill*. These are names for certain ratios. When you make cookies or mix mortar, and you *double a recipe*, you are using ratios. Figure 9-1 shows some situations where ratios are used.

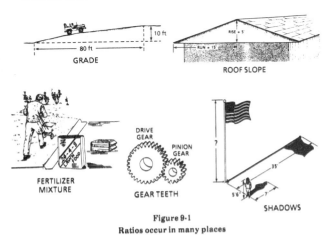

Figure 9-1
Ratios occur in many places

When you say that two ratios are equal, you have a **proportion**. You can use proportions to find a quantity you can't easily measure. For example, you can figure out the height of a flagpole without measuring it by using a proportion. Ratios and proportions make

- The mathematical concept of *ratio and proportion* is both useful and important, especially in tasks that involve recipes, scaling, and preparing/reading blueprints and scale drawings. Understanding this skill will enable students to compare quantities numerically, to form proportions, to solve for unknowns, and to generalize relationships. Your students will use the basic concepts involved in ratio and proportion to solve practical problems. Being able to recognize valid proportions, direct or indirect, and reason that "if a is to b, then c is to d" is a powerful step forward in problem solving.
- Use Figure 9-1 to acquaint students with ratios and the many common places where they are found and used. The figure shows ratios that specify the *grade of a hill*, the *pitch or slope of a roof*, the proportions of *chemical elements in a fertilizer mixture*, the number of *gear teeth in meshing gears* and the *comparison of shadow lengths*. Ask your students to think of examples where they use ratios—or where they know ratios are used. Also, ask them to recall instances where ratios were used in the video.

Using Ratios and Proportions
Page 4

possible the use of scale models to design buildings, cars, and airplanes.

As you watch the video, notice when people compare quantities to write a ratio, and when they use equal ratios (a proportion) to solve problems.

IDENTIFYING RATIOS

A ratio is a way to compare two numbers, or to compare the measures of two quantities. Let's see how that's done.

What do ratios do?

A ratio *compares* one number to another. For example, you mix one pint of oil with two gallons of gasoline for a certain lawn mower. Since there are 8 pints to a gallon, this is a ratio of 1 part oil to 16 parts gasoline.

You can use a ratio to compare two measurements for a circle. Suppose that you have several circular objects such as a saucer, a jar, a fifty-cent coin, a wheel, and a phonograph record. You can measure the diameter and circumference of each circle and write down all your measurements in a table.

With the measurements in your table, you can then *compare* the circumference and the diameter of a wheel, even though you haven't measured it. You can write this comparison as a fraction that shows the circumference divided by the diameter of the wheel. For example:

$$\frac{\text{Circumference of wheel}}{\text{Diameter of wheel}} = 3.1416$$

You may already know that this particular fraction always results in the *same* number. We call that number, whose value is about $^{22}/_7$ or 3.1416, π or pi. This fact is also expressed in the formula for the circumference of a circle:

$$\text{Circumference} = \pi \times \text{diameter}$$

or

$$C = \pi d$$

The **comparison** of circumference and diameter is a **ratio**. You can say "For any circle, the *ratio* of the circumference *to* the diameter is

Applied Mathematics

- Although a ratio is often expressed as a fraction, and it may be said that a fraction is a ratio, the two are not quite identical. You may or may not wish to discuss some aspects of this question with your class.

  For example, the ratio of males to females in a class in a boys' school might be 23 to 0, a perfectly good ratio. Or the score of a basketball game might, at some point, be 6 to 0, another ratio. However, when we try to express either of these ratios as fractions, we run into trouble. The ratio $^{23}/_0$—since division by zero is not defined—is an *undefined quantity*. Although it can be written in the form of a fraction, it is **meaningless** as a number. However, the expression 23 to 0 or 23 : 0 remains a meaningful and useful form for a ratio.

- This discussion may be too abstract for some classes but, if you choose to get into it, one way to show that $^{23}/_0$ is meaningless as a number is to establish that $^{10}/_5 = 2$ can be checked by multiplying 5 by 2 to give 10. When this same method is tried with $^{23}/_0 = y$, one cannot find a number for y that will make 0 times y give 23.

pi." Notice that when you compare C to d, the quantity that comes first (C) is the numerator of the fraction, and the second quantity (d) is the denominator.

Write a fraction that shows the ratio of the diameter of a circle to its radius (using d for diameter and r for radius).

 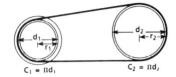

Figure 9-2
Comparing circle measures

For any circle, the fraction C/d always equals pi. What does the fraction d/r always equal? Use the picture in Figure 9-2 to help you.

Write the fraction that shows the ratio of r to d. What does this fraction always equal?

Does it make any difference whether you say "the ratio of d to r" or "the ratio of r to d"?

Study Activity: Use a ratio to compare the number of females to the number of males in your class. Now write a ratio that compares the number of males to the number of females. Are these two ratios different? Figure 9-3 shows one possibility.

Figure 9-3
Comparing other quantities

- The fraction showing the ratio of a circle's diameter to its radius is d/r. The fraction d/r is always equal to 2. The fraction r/d is always equal to $^1/_2$. Therefore, it does make a difference whether one says "the ratio of d to r" or "the ratio of r to d."
- The ratios $^{\text{\# females}}/_{\text{\# males}}$ and $^{\text{\# males}}/_{\text{\# females}}$ will be different for each class except, of course, for the situation where there are equal numbers of each. In Figure 9-3, there are 13 students—5 boys and 8 girls. To help your students verbalize the words in a ratio, write the ratios $^5/_{13}$, $^8/_{13}$, $^5/_8$ and $^8/_5$ on the chalkboard. Then ask your students to interpret each of these ratios in terms of what's shown in Figure 9-3.

- You may want to use a right triangle and have students write ratios of the triangle's sides—a = altitude or height, b = base, c = hypotenuse—as follows:

a/b, b/a, a/c, c/a, b/c and c/b.

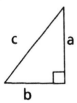

This exercise can serve as an "awareness-level" introduction to the trigonometric functions of a right triangle.

Write a ratio that compares the number of females in your class to the total number of people in your class.

What kinds of ratios are there?

When you wrote the ratio of females to people, you compared a part of the class to the whole class. Many ratios compare a part to a whole.

Other ratios compare a part to another part. You wrote one of these when you compared the number of females to the number of males.

Ratios that have a constant value. Some ratios can be written as a fraction that always has the same value. For example, the ratio of a circle's diameter to its radius is always equal to two. These ratios, whose value is always the same, are called **constant** ratios.

Write another ratio related to a circle that is a constant ratio. Is the ratio of males to females the same in every class? Is this a constant ratio?

The scores for a game can also be a ratio. "We beat the home team thirteen to seven!" Are game scores constant ratios?

Using units with ratios. When a ratio compares two measures, what happens to the units? In the d/r ratio for a circle, the diameter and the radius are always measured in the same units. For example, consider this ratio of diameter to radius:

$$\frac{d}{r} = \frac{12 \text{ in.}}{6 \text{ in.}} = \left(\frac{12}{6}\right)\left(\frac{\text{in.}}{\text{in.}}\right)$$

Since you can treat the fraction of units—$^{\text{inches}}/_{\text{inches}}$ as 1, this fraction becomes simply

$$\frac{12}{6} \times 1 = 2 \times 1 = 2$$

When both terms (the numerator and the denominator) of a ratio have the *same* units, you can write the ratio without the units. For example, in the score for a game, you can say "13 to 7" without saying "points scored."

When both terms of a ratio have the **same units**, the ratio is called a **proper ratio**. Proper ratios can be written without the units.

Applied Mathematics

- There are many ways to classify ratios. Only a few are suggested here. The main point here is to enable the students to use ratios to solve problems. Therefore, naming and classifying are kept to a minimum.
- The ratio c/d or $c/2r$ is a constant ratio for a circle. Game scores are not constant ratios.
- If teachers want to expand the constant-ratio concept, several 30°–60°–90° triangles with different dimensions can be constructed, as shown. Comparisons of the ratios of corresponding sides—**by measurement**—should convince students that these ratios have the same values. All six ratios can be compared. For example, measure of a/c = measure of d/f, measure of c/a = measure of f/d, and so on.

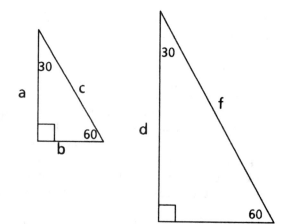

(Teacher notes continued on page 28a.)

In the ratio that compares the number of females and males, you could write this fraction:

$$\frac{8 \text{ females}}{5 \text{ males}}$$

and the units are *not* the same so they don't "cancel." You can read this fraction as "The ratio of females to males is 8 to 5." You might also say, "The ratio is 8 females to 5 males."

Ratios often compare measurements with unlike units. **If a ratio is used to compare measures with different units, the units must be given in the ratio.**

Comparing distance with time. Suppose a car travels 100 miles in 2 hours. We can write the ratio that compares **miles traveled** to **time spent**. If this ratio is written as a fraction, and the fraction is simplified—by dividing numerator and denominator by 2—the result is this:

$$\frac{100 \text{ mi}}{2 \text{ hr}} = \frac{100 \text{ mi} \div 2}{2 \text{ hr} \div 2} = \frac{50 \text{ mi}}{1 \text{ hr}} = 50 \text{ mi per hr or } 50 \text{ mph}$$

There's that word *per* again! Recall that in *percent*, **per** means **for each** (or *divided by*) 100. Here it means **for each** (or *divided by*) one hour.

Now you know at least three ways to write a ratio—as a fraction, a division, or in words, using **to** or **per**.

Many ratios are also known as **rates**—a word closely related to ratio. Ratios that compare a **distance traveled** to **time spent**, such as "50 miles per hour," are rates. You may encounter other rates such as a sale price of 5 cans for $1, a mileage allowance of 22¢ per mile, or a wage of $5.25 per hour.

Study Activity: Write a ratio that tells how fast a messenger delivers packages with a bicycle if the messenger averages 10 miles in two hours.

- The vocabulary usage for the term **ratio** is not precise. Some authorities define a *ratio* as having the same units and a *rate* as having different units. Others see this as an unnecessarily fine distinction and follow common English usage by calling any comparison a ratio, as we do here. For example, although we can speak of the "rate of females to males," it is much more common to call this a ratio. Others would reserve the term rate for ratios that involve time. In any case, the important point is that units should be included if they are not identical.
- If your students have some difficulty with units, you may want to show several examples of ratios that involve units:

▸ Proper ratios that have the same units, like these:

$$\frac{6 \text{ in}}{12 \text{ in}} \; ; \; \frac{3 \text{ cm}^2}{13 \text{ cm}^2} \; ; \; \frac{6 \text{ oz}}{18 \text{ oz}} \text{, etc.}$$

▸ Other ratios that have different units, like these:

$$\frac{1.5 \text{ grams}}{\text{liter}} \; ; \; \frac{3 \text{ revolutions}}{2 \text{ seconds}} \; ; \; \frac{60 \text{ miles}}{\text{hour}} \text{, etc.}$$

(Teachers notes continued on page 28a.)

READING AND INTERPRETING RATIOS

You know several ways to write a ratio—as a fraction, a division, or in words, using *to* or *per*. There is one more common way to write a ratio. For example, concrete may be mixed using one part of cement to two parts of sand. This ratio is often written as 1 : 2, cement to sand. Notice that the ratio 1 : 2 doesn't tell us which is cement and which is sand. *Words* go with the ratio so that you can tell which part of the mixture is which.

How do you write ratios?

Write the comparison of one part cement to two parts sand in four different ways.

When bricklayers use mortar, they may have more than just two ingredients. A colon (:) is used to write these comparisons too. For example, a mortar for an outdoor brick wall might have 1 part cement, 1 part lime, and 6 parts sand. This is written 1 : 1 : 6—cement, lime, and sand. You read this as "one to one to six."

Ratios with three terms are also used on the labels of lawn food. In this case, the numbers are always in the same order—the first number shows the percentage of nitrogen, the second number shows the percentage of phosphorus, and the third number shows the percentage of potassium. Figure 9-4 shows lawn food with a 24-4-8 mixture.

Figure 9-4
Ratios for lawn food

Study Activity: In a fertilizer labeled 25-3-5, what is the ratio of nitrogen to potassium?

- Answers to the question on four different ways to compare concrete mixtures are:
    - 1 : 2 cement to sand
    - $\frac{1}{2}$
    - 1 to 2
    - 1 part cement per 2 parts sand
- You should not spend too much time with 3-part mixtures since the primary goal is to build on student understanding of proportions for which the equality between two-term ratios is used.
- In writing ratios, practice with your students to be sure that there is a good understanding of what the numerator is being compared to—or with. You may want to use a practice sheet so that students gain experience in writing ratios with the correct numerator and denominator. Definitions of terms your students use every day can also be practiced. A few examples are suggested for your consideration:
    ▶ mph (speed) is the ratio of distance to time
    ▶ rpm (rotational speed) is the ratio of revolutions to time

▶ $\text{Average} = \dfrac{\text{total quantity}}{\text{number of examples}}$

▶ $\text{Profit} = \dfrac{\text{gain}}{\text{cost}}$

(Teachers notes continued on next page.)

Using Ratios and Proportions
Page 9

What does a ratio tell you?

A ratio tells you how one quantity compares with another. A ratio uses numbers (sometimes with units) to make the comparison. For instance, ratios are used to describe the strength of solutions used by health-care workers. A solution is a liquid preparation of a medicine (chemical compound) dissolved in a liquid, usually water.

If there are 3 parts of a saturated solution of boric acid in 12 parts of water, how can you find the ratio of boric acid solution to water?

You can find the ratio by writing a fraction that compares the amount of solution to the amount of water:

$$\frac{3 \text{ parts boric acid solution}}{12 \text{ parts water}}$$

In some instances, you may want to simplify the fraction. If so, the ratio of boric acid solution to water becomes $\frac{1}{4}$ or 1 to 4. So, you can say either "3 to 12" or "1 to 4."

In an experiment to test the effect of light on the plant growth rate, plants in tray A grew 3 inches in the same time that plants in tray B grew 0.75 foot. What is the ratio of the growth for plants in tray A compared to plants in tray B?

Begin by writing the fraction that compares the two growths. Include the units and use the same order for the terms as in the question:

$$\frac{\text{Plant growth in tray A}}{\text{Plant growth in tray B}} = \frac{3 \text{ in.}}{0.75 \text{ ft}}$$

This time the units are *not* the same. Change the units to the same form before you simplify the fraction. You can choose either inches or feet. The fraction on the left shows inches; the one on the right shows feet.

$$\frac{3 \text{ in.}}{9 \text{ in.}} \qquad \frac{0.25 \text{ ft}}{0.75 \text{ ft}}$$

Now simplify the fractions.

$$\frac{3 \text{ in.}}{9 \text{ in.}} = \frac{3 \text{ in.} \div 3}{9 \text{ in.} \div 3} = \frac{1}{3} \qquad \frac{0.25 \text{ ft}}{0.75 \text{ ft}} = \frac{0.25 \text{ ft} \div 0.25}{0.75 \text{ ft} \div 0.25} = \frac{1}{3}$$

Whether you work the problem in inches or feet, the final ratio of growth in tray A compared to tray B is 1 to 3.

Continued from page 8
- The mixture ratio 24-4-8 in Figure 9-4 can also be shown as 6-1-2.
- *Answer to the Study Activity:*
  In a fertilizer labeled 25-3-5, the ratio of nitrogen to potassium is 25 to 5 or 5 to 1.

....

- The comparison of saturated boric acid solution to water is typical of a *part-to-part* comparison. Such comparisons are common in the preparation of recipes, where different ingredients are mixed in the batter; in the preparation of solid mixtures where sand, cement and gravel are mixed to make concrete; and in the preparation of fertilizers where different parts of chemical elements are mixed to give the right balance of nutrients.
- A comparison of different quantities—*height of growth* and *time*—is illustrated in the experiment that describes the effect of light on the growth rate of plants. This ratio, quite different from the "part-to-part" comparison, provides another useful example of how ratios are used.
- You may want to have your students study other examples of "part-to-part" and "part-to-whole." You may also want to review percentage problems as parts per one hundred, or compare the relationship of metric units to each other. All are valid comparisons that involve ratios.

COMPARING RATIOS

You have learned that a ratio compares two quantities. Sometimes it is useful to compare two ratios. For example, if one mixture of liquid insecticide and water has a ratio of 1 : 3 and another has a ratio of 3 : 9, are the two mixtures equal in strength? The next section explains how to compare ratios.

When are ratios equal?

You can probably see right away that a mixture with a ratio of 3 : 9 is equal to one with a ratio of 1 : 3 because the fraction $3/9$ simplifies to $1/3$.

Sometimes it is not as easy to tell whether two ratios are equal. For example, is the ratio 15 : 10 equal to the ratio 21 : 14? One way to compare ratios is to express both as a fraction, simplify each fraction, and then compare them.

If the simplest forms of the fractions for two ratios are equal, then the ratios are equal.

Find the simplified forms of $15/10$ and $21/14$ to see if they are equal.

Since the fractions are each equal to $3/2$, in simplest form, you can equate them like this:

$$\frac{15}{10} = \frac{21}{14}$$

Of course, you can always use your calculator to tell if two ratios are equal. For example, you can enter the ratio $15/10$ (or 15 ÷ 10) and get the answer 1.5. Then you can enter the ratio $21/14$ (or 21 ÷ 14) and get the same answer—1.5. Since the answers are identical, the ratios are equal.

Another way to compare ratios.

Here's another way to check whether or not two ratios are equal. Look at Figure 9-5. In this figure, two ratios—a/b and c/d—are compared, and the question "Are the two ratios equal?" is answered.

Applied Mathematics

Using Ratios and Proportions
Page 10

- The answer to the question—Is 15 : 10 equal to 21 : 14?—is yes. You may want to have your students simplify each fraction to show the identity $3/2 = 3/2$. The expression $15/10 = 21/14$ is a "linkage" from ratios to proportions. You may want to support the linkage by using equivalent fractions to construct a few other proportions, such as $1/2 = 4/8$, $1/3 = 3/9$, $1/5 = 2/10$.
- The *fair test* for determining if a proportion is properly constructed is to simplify fractions on both sides of the proportion to see if an identity (identical expression) can be obtained. An example is shown below:

$$\text{Does } \frac{16}{24} = \frac{48}{72} \text{ ?}$$

Simplify: $\dfrac{2\,(8)}{3\,(8)} = \dfrac{2\,(24)}{3\,(24)}$

The result is: $\dfrac{2}{3} = \dfrac{2}{3}$, an identity.

Therefore, $\dfrac{16}{24} = \dfrac{48}{72}$.

Using Ratios and Proportions
Page 11

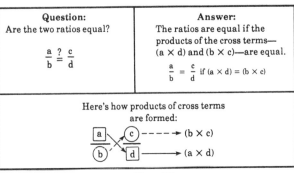

Question:	Answer:
Are the two ratios equal? $\dfrac{a}{b} \stackrel{?}{=} \dfrac{c}{d}$	The ratios are equal if the products of the cross terms—$(a \times d)$ and $(b \times c)$—are equal. $\dfrac{a}{b} = \dfrac{c}{d}$ if $(a \times d) = (b \times c)$

Here's how products of cross terms are formed:

Figure 9-5
Comparing ratios

As explained in Figure 9-5, the ratios are equal if the two products of the cross terms—$(a \times d)$ and $(b \times c)$—are equal. The cross terms are formed by multiplying the upper left term by the lower right term $(a \times d)$—and the lower left term by the upper right term $(b \times c)$.

Let's use the method outlined in Figure 9-5 to check whether the two ratios $\dfrac{25}{10}$ and $\dfrac{35}{14}$ are equal.

- First, write the two ratios as an equality with a question mark over the equal sign. (You don't know yet if the two fractions are really equal.)

$$\dfrac{25}{10} \stackrel{?}{=} \dfrac{35}{14}$$

- Then find the product of the cross terms and check to see if the two products are equal.

$$25 \times 14 \stackrel{?}{=} 10 \times 35$$
$$350 = 350 \text{ (products \underline{are} equal)}$$

Since the products of the cross terms are equal, the ratio $25/10$ equals the ratio $35/14$. So you *can* write the equality $25/10 = 35/14$.

The equality—or equation—$\dfrac{25}{10} = \dfrac{35}{14}$ is a valid proportion.

Study Activity: Use either of the methods you've learned to decide whether these two liquid solutions have equal ratios of alcohol and water.

- Take a few minutes to explain the paths and arrows in Figure 9-5 that illustrate how to multiply the cross terms in a proportion to get the equal products. Do several examples with numbers (equivalent fractions) so that your students who are not comfortable with letters (a,b,c,d) get the basic message of how to multiply the cross terms to form the equal products.
- The diagram in Figure 9-5 is intended to give a visual definition of how equal products of cross terms are formed. You may want to show students how the equal products are obtained mathematically, as follows:

  ▸ Given $\dfrac{a}{b} = \dfrac{c}{d}$

  ▸ multiply by bd; then $\dfrac{a}{b}(bd) = \dfrac{c}{d}(bd)$

  ▸ $\dfrac{a(\cancel{b}d)}{\cancel{b}} = \dfrac{c(b\cancel{d})}{\cancel{d}}$, so $ad = cb$.

- **CAUTION:** Students who have learned how to multiply fractions—**numer $\times$ numer** and **denom $\times$ denom**—can easily become confused when the products of cross terms—**numer $\times$ denom** and **denom $\times$ numer**—are formed to check a proportion. Take a few minutes to distinguish between the two operations—and head off any possible confusion.

One solution has a ratio of 55 parts alcohol to 20 parts water. The second has a ratio of 33 parts alcohol to 12 parts water.

Equal ratios can be helpful in drawing diagrams, such as the ones in Figure 9-6. Each of the triangles shown has the same shape, but each is of a different size.

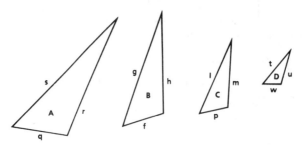

Figure 9-6
Similar triangles: same shape—different sizes

What are similar figures?

Similar figures may have different sizes, but they will have the same shape. All of the triangles in Figure 9-6 are **similar**.

You could take the smallest triangle and **enlarge** it—with a photocopying machine or a photographic enlarger so that all sides are increased in proportion—and make it the same size as the other triangles. Only the size changes, not the shape. Or, you could take the largest triangle and reduce it **proportionally** to get the smaller ones.

If two figures are similar, the ratios of the **corresponding (or matching) sides** are the same. In Figure 9-6, for example, the ratio of the longest side of triangle C to the longest side of triangle D is equal to the ratio of the shortest side of triangle C to the shortest side of triangle D. In terms of the letters shown there, this would be:

$$\frac{l}{t} = \frac{p}{w}$$

12   Applied Mathematics

- *Answer to the Study Activity*:
  Your students should show that

  $$\frac{55 \text{ parts alcohol}}{20 \text{ parts water}} = \frac{33 \text{ parts alcohol}}{12 \text{ parts water}}$$

  by proving that $^{55}/_{20}$ and $^{33}/_{12}$ are equivalent fractions, each equal to $^{11}/_{4}$ in simplified form. They can also form products of the cross terms—ignoring the units—and show that the products are equal, that is, $55 \times 12 = 20 \times 33$, each equal to 660.
- The triangles shown in Figure 9-6 are **similar** and have been drawn with corresponding sides proportional in length. Therefore, students can use a centimeter scale and measure the matching sides (to within a millimeter) to check out the equal ratios.
- Your students may not understand what is meant by "corresponding sides of similar figures." Therefore, you should draw similar triangles or similar rectangles and practice with them. It may be worth some time for class activities to contrast **similarity** (same shape and proportional parts) with **congruency** (same shape and size). Also, you may want to use Figure 9-6 to point out that corresponding sides of two similar triangles are always opposite equal angles.

(*Teacher notes continued on page 28a.*)

Study Activity: If, in Figure 9-6, the ratio of side f in triangle B to side q in triangle A is 3 to 4 (and the two triangles are similar), what is the ratio of side g to side s?

Now look at Figure 9-7. Which of the rectangles shown there are similar? Ratios can help you answer that question. To check out the ratios more easily, the sides have been labeled ℓ for length and w for width.

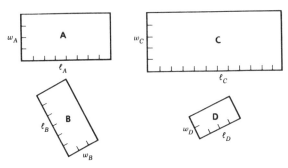

Figure 9-7
Similar rectangles?

If the width w_A of rectangle A is 4 units, and the width w_B of rectangle B is 3 units, what is the ratio of the width of rectangle A to rectangle B? If the length ℓ_A of rectangle A is 8 units and the length ℓ_B of rectangle B is 6 units, what is the ratio of the lengths of rectangle A to rectangle B? Are these two ratios equal? If the two ratios are equal, then the rectangles are similar. That is, if,

$$\frac{w_A}{w_B} = \frac{\ell_A}{\ell_B}$$

then rectangle A is similar to rectangle B.

Examine the dimensions of rectangle B and rectangle C. Use this same method to test whether rectangles B and C are similar. What is your conclusion? Examine rectangle D. Is rectangle D similar to any other rectangle in Figure 9-7?

- *Solution to the Study Activity:*
  If the ratio of side f in triangle B to side q in triangle A is 3 to 4, then the ratio of side g in triangle B to side s in triangle A is also 3 to 4. Sides f and q are corresponding sides, as are sides g and s.
- Regarding the similarity of rectangles in Figure 9-7, they have been drawn so that A, B and D are similar to each other—with width-to-length in the ratio of 1 : 2. Rectangle C is not similar to any of the others.

Answer to questions about rectangles in Figure 9-7.

▶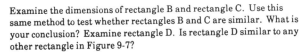
  The ratios are equal.

▶ B and C are not similar. For example,
  $$\frac{w_B}{w_C} = \frac{3}{5} \; ; \; \frac{\ell_B}{\ell_C} = \frac{6}{12} \; ; \; \frac{3}{5} \neq \frac{6}{12}$$

▶ Rectangle D is similar to A and B.
▶ Your students can count the unit lengths in each figure to form ratios of corresponding sides. Thus, they can show that A, B and D are similar to each other, and that C is not similar to the others.

Using Ratios and Proportions
Page 14

Suppose you have a pattern for a rectangularly shaped tabletop. The pattern is 3" wide and 4" long and you want to enlarge this pattern to make a table that is 5' wide. You can draw a rectangle (on a sheet of plywood) that is **similar** to the rectangle in your pattern. The ratio of the width of the pattern to the width of the table is 3" to 5', or 3" to 60", or in simplified form, 1 to 20. The ratio of the lengths will also be 1 to 20. How long will the table be? The rest of this unit helps you find the answer to that question—and others like it.

IDENTIFYING AND WRITING PROPORTIONS

An expression that equates one ratio to another is called a **proportion**. Here is an example of a proportion:

$$\frac{2}{3} = \frac{4}{6}$$

How can you identify a proportion?

An expression written in this form is a proportion only if the first ratio equals the second ratio. For example, the following statement tells you that you **do not** have a proportion.

$$\frac{2}{3} \neq \frac{3}{4}$$

You know that, of course, because the fraction ²/₃ does not equal the fraction ³/₄.

To find out whether expressions like these are true proportions, you could try the test for products of cross terms (shown in Figure 9-5). If the products of the cross terms are equal, the two ratios are equal and the expression is a proportion.

The corresponding sides of similar figures are always proportional. You can always write a proportion saying that the ratio of two sides in one of two similar figures is the same as the ratio for the *corresponding sides* in the other.

Proportional figures are used whenever someone follows a pattern or model. A builder, for example, may have plans or drawings showing a ceiling height of 8 inches. But the ceiling built from those plans is

Applied Mathematics

- Draw a sketch of the tabletop problem on the chalkboard to help students visualize the general problem. (See drawing at right.) Use the problem as a motivator for using ratios and proportions. Do not attempt to solve it at this point. However, you can identify the **corresponding sides** and set up the proportion. Note that we are using the letter "y" for the unknown length of the tabletop

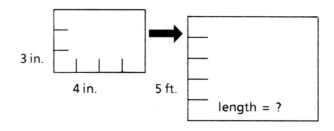

The proportion is: $\dfrac{3 \text{ inches}}{4 \text{ inches}} = \dfrac{5 \text{ feet}}{y \text{ feet}}$

Later we will solve this proportion to find y, the desired length of the tabletop.

Applied Mathematics

Using Ratios and Proportions
Page 15

8 feet high. The *scale* for that plan is 1 inch = 1 foot. Notice that in this scale the units are not the same.

Write the ratio (or scale) for those plans using the *same unit* for the plan and the building.

From this ratio, you can write this proportion:

$$\frac{1 \text{ in. on the plan}}{12 \text{ in. for the building}} = \frac{8 \text{ in. on the plan}}{8 \times 12 \text{ in. for the building}}$$

Study Activity: Use this model to write a proportion showing that a countertop drawn to scale that is 3 inches high will actually be 36 inches high in the building.

How can you write a proportion?

When you set up a proportion, you write an expression that involves the comparison of certain measurements. To avoid making careless errors in writing a proportion, try writing the proportion in words first.

For example:

$$\frac{\text{Width of rectangle A}}{\text{Width of rectangle B}} = \frac{\text{Length of rectangle A}}{\text{Length of rectangle B}}$$

Notice that the ratio on the left is set up the same way as the ratio on the right. The left ratio compares width A to width B and the right ratio compares the *corresponding* lengths—length A to length B.

Another comparison may be:

$$\frac{\text{Width of rectangle A}}{\text{Length of rectangle A}} = \frac{\text{Width of rectangle B}}{\text{Length of rectangle B}}$$

In this proportion, the comparison changes, but it is still the *same* comparison in both ratios—width A to length A and width B to length B.

What will NOT give a proportion is a certain comparison in the first ratio and a DIFFERENT comparison in the second ratio. For example,

$$\frac{\text{Width A}}{\text{Width B}} \neq \frac{\text{Length B}}{\text{Length A}}$$

- **Answers to the Study Activity:**

$$\frac{1 \text{ in. } (plan)}{12 \text{ in. } (building)} = \frac{3 \text{ in. } (plan)}{(3 \times 12 \text{ in.}) (countertop)}$$

Thus $\frac{1 \text{ in.}}{12 \text{ in.}} = \frac{3 \text{ in.}}{36 \text{ in.}}$; *a valid proportion*

- You may find it useful to conduct a class activity with similar triangles and practice writing proportions. (See the figures drawn at the right.)

(Teacher Notes continued on page 28a.)

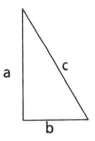
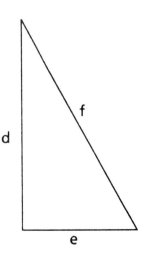

Notice that this is *not* a proportion because the first ratio compares a measure from A to a measure from B. But the second ratio reverses this comparison.

Let's return to the tabletop problem. The pattern for the tabletop measures 3" by 4" and you want to enlarge this to make a table that is 5' wide. How long will the table be?

For this problem, you could write several different proportions. Here is one of the possibilities (first in words and then with the numbers):

$$\frac{\text{Width on pattern}}{\text{Length on pattern}} = \frac{\text{Width of the tabletop}}{\text{Length of the tabletop}}$$

$$\frac{3 \text{ in. on pattern}}{4 \text{ in. on pattern}} = \frac{5 \text{ ft for tabletop}}{? \text{ ft for tabletop}}$$

The next section shows you how to find the ? in this problem, and how to solve other similar problems.

SOLVING PROPORTIONS

Suppose you have a proportion that is missing one term—such as in the following case:

$$\frac{3}{5} = \frac{y}{10}$$

In this proportion, we've let the letter y (instead of the symbol ?) stand for the missing term—or the unknown value. The problem now becomes—"What value must y have to make this a true proportion?" Let's look at several ways to figure out what y must be.

Finding the unknown term

Since this proportion $3/5 = y/10$ involves simple numbers, try solving for y this way. Ask yourself; "What can I do to the 5 (the denominator of one fraction) to get 10 (the denominator of the other fraction)?" You would multiply by 2. Then do the same thing to the number 3 to get the value of y. That would give you $2 \times 3 = 6$. So the value for y must be 6. This, then, results in the following proportion:

$$\frac{3}{5} = \frac{6}{10}$$

Applied Mathematics

Using Ratios and Proportions Page 16

- The methodology comparing the widths and lengths of rectangles A and B intends to stress the need to be careful of the matching order for the terms in a proportion. Be sure to point out that in almost every situation, more than one valid proportion that leads to the correct solution of a missing term can be written. Point out also—as does the text—that all ratios that can be written may not be equated to form a valid proportion. That was the point of the note on the previous page.
- You may want to give students some freedom to write a proportion that makes sense to them. For example, if they are finding one of the missing sides for the two similar triangles shown, they might proceed as follows:

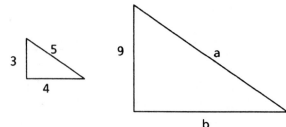

Any one of the following proportions will determine b:

$$\frac{4}{3} = \frac{b}{9} \; ; \; \frac{4}{b} = \frac{3}{9} \; ; \; \frac{3}{4} = \frac{9}{b} \; ; \; \frac{b}{4} = \frac{9}{3} \; ; \; b = 12$$

(Teacher notes continued on page 28a.)

Applied Mathematics

Using Ratios and Proportions Page 17

Is this a true proportion? Are the two ratios equal to each other?

Here's another method you can use—even when the numbers are not simple. You know that for a true proportion, the products of the cross terms must be equal. For the proportion $3/5 = y/10$, then, the products of the cross terms are 3×10 and $5 \times y$—and they must be equal. Equate them to get:

$$5 \times y = 3 \times 10$$
$$5y = 30$$

What value of y satisfies the equality $5y = 30$? Do you see that only $y = 6$ makes this a true equality? How would you solve the equation $5y = 30$ to show that y must equal 6?

Here's a general way to solve the equation $5y = 30$ to get $y = 6$.

Start with the equation:

$$5y = 30$$

Next, divide each side by whatever number multiplies the unknown value y—in this case, the number 5. This operation isolates y—leaves y standing by itself—as follows:

$$\frac{5y}{5} = \frac{30}{5} \qquad \text{(The fives on the left divide to give 1.)}$$

$$1y = \frac{30}{5} \qquad \text{(You can omit the 1, since } 1y = y.)$$

$$y = \frac{30}{5} \qquad \text{(Divide 30 by 5 to get 6.)}$$

$$y = 6$$

Try the general method of equating products of the cross terms to find the unknown in the following proportion:

$$\frac{15}{21} = \frac{y}{7}$$

First equate the two products of cross terms:

$$21 \times y = 15 \times 7, \text{ or}$$
$$21y = 105$$

- Most students will probably be able to solve the simple equation $5y = 30$ "in their heads." Nevertheless, we introduce the general method for isolating the *unknown* by using the appropriate arithmetic operation—here the *division of equals by equals*. Spend some time going over the examples.
- We have examined two proportions on this page—$3/5 = y/10$ and $15/21 = y/7$—both with the unknown y in the numerator of a ratio. It may be worthwhile to write other proportions, with the unknown in the denominator of one of the ratios, and also with different letters, just to broaden students' thinking patterns. For example, write proportions such as $1/2 = 4/z$ and $3/27 = 9/p$, etc. Show them that forming the products of cross terms in these proportions also leads to a simple equation, which then can be solved for the unknown by the method outlined on this page.

Next, free, or isolate, the y on the left side of the equation by dividing each side of the equation by 21.

$$\frac{\cancel{21}y}{\cancel{21}} = \frac{105}{21} \qquad \text{(Divide 21 out on the left to give 1.)}$$

$$y = \frac{105}{21} \qquad (\frac{105}{21} = 105 \div 21 = 5)$$

Therefore, $y = 5$.

So, in the proportion $^{105}/_{21} = {}^y/_7$, you have shown that the missing number (y) must be equal to 5.

Example 1:
Using proportions in recipes

Now follow through this example:

A recipe for making cookies uses 4 cups of flour and produces 32 cookies. How many cups of flour should you use if you want to produce 4 dozen cookies?

First, write a proportion in words to make sure you use the same comparison for both ratios. Begin with what you are looking for as the first term:

$$\frac{\text{Flour you want to use}}{\text{Cookies you want}} = \frac{\text{Flour in the recipe}}{\text{Cookies in the recipe}}$$

Then put in the numbers you know, using the letter y for the unknown value to be found:

$$\frac{y \text{ cups of flour}}{4 \text{ dozen cookies}} = \frac{4 \text{ cups}}{32 \text{ cookies}}$$

Convert units so that they are the same in the matching terms of the comparison:

$$\frac{y \text{ cups}}{48 \text{ cookies}} = \frac{4 \text{ cups}}{32 \text{ cookies}}$$

Since the units are the same in the matching terms, you can now write the proportion simply as:

$$\frac{y}{48} = \frac{4}{32}$$

To solve for y, first write the equality between the products of the cross terms:

Applied Mathematics

- Work through Example 1. It is a problem that may be familiar to many of your students. Emphasize that two steps are most important here:
  ▸ Set up the proportion to make sure that you use the *same comparison* for *both* ratios:

  $$\frac{\text{flour used}}{\text{cookies to be made}}$$

  ▸ Convert units so that they are the same for the *matching terms* in the two ratios. Point out, for example that the units in the ratio

  $$\frac{\text{cups of flour}}{\text{dozens of cookies}}$$

  do not match the units in the ratio

  $$\frac{\text{cups}}{\text{cookies}},$$

  therefore *dozens of cookies* have to be converted to *cookies*.

Using Ratios and Proportions
Page 19

$$32y = 4 \times 48$$
or $32y = 192$

Then isolate y and solve:

$$\frac{\cancel{32}y}{\cancel{32}} = \frac{192}{32}$$

$$y = \frac{192}{32}$$

$$y = 6$$

The answer, then, is that you will need 6 cups of flour to make 4 dozen cookies.

Ask yourself if this is a reasonable answer. To check, write the proportion and find the products of the cross terms (or simplify each fraction) and see if they are really equal:

$$\frac{6}{48} = \frac{4}{32}$$

$$6 \times 32 = 192 \quad \text{and} \quad 4 \times 48 = 192$$

Apply the rule involving products of cross terms to solve these problems:

Study Activity:

1. One family uses an average of 400 gallons of water per day. How many families will a supply of 6840 gallons serve for a day?

2. A basketball team makes 7 out of 11 of their attempted free throws in the first game of the season. If they continue to make free throws at this rate, how many free throws would the team be expected to make in 187 tries?

Let's summarize:

First, write a proportion in words to make sure you use the same comparison for both ratios.

Then put in the numbers you know, letting a letter such as "y" stand for the unknown number.

Convert units so that they are the same in the matching terms of the comparison.

- Continue to emphasize the importance of checking results. Point out that checking the answer for an unknown, found by solving a proportion, is done easily. The proportion is set up again, with the number found for the unknown in its proper place, then the products of the cross terms are formed and checked for equality.
- *Answers to the Study Activity:*

$$\frac{1 \text{ family}}{400 \text{ gallons}} = \frac{(y) \text{ families}}{6,840 \text{ gallons}}$$

y is 17.1, or about 17 families.

$$\frac{7 \text{ free throws made}}{11 \text{ free throws tried}} = \frac{(y) \text{ free throws made}}{187 \text{ tried}}$$

The answer y is 119 free throws.

- The *Examples* and the *Study Activities* give you an opportunity to stress some comparison techniques:
  ▸ You check the **order** to see if cups/cookies are compared to cups/cookies, if families/gallons is compared to families/gallons and if free throws made/free throws tried is compared to free throws made/free throws tried. You also check to determine which way you are scaling (up or down) to determine whether the answer is reasonable.
  ▸ You check your answer by determining if the products of the cross terms are identical (this is the "fair test" for any proportion).

Using Ratios and Proportions
Page 20

To find the unknown value y, equate the products of the cross terms to get an equation that involves y. Then solve the equation for y.

Ask yourself if your answer for y is a reasonable answer. To check, write the proportion and find the cross products of the cross terms to see if they are really equal.

Study Activity: Now you can easily solve the tabletop problem. The rectangular-shaped pattern for the tabletop measures 3" by 4" and you want to make a table that is 5' wide. How *long* will the table be?

$$\frac{\text{Length of pattern}}{\text{Width of pattern}} = \frac{\text{Unknown length of the tabletop}}{\text{Width of the tabletop}}$$

$$\frac{4\text{" on the pattern}}{3\text{" on the pattern}} = \frac{y \text{ for tabletop}}{5' \text{ for tabletop}}$$

Find the length (y) of the tabletop. Write the correct units for y and check your answer.

Applying ratios and proportions to problems

Proportions are often used with patterns, with drawings, for furniture, with blueprints for buildings, and for maps.

On one highway map a distance of 1 inch represents a distance of 100 miles. What is the actual distance between two towns if they are separated by 4.25 inches on the map? Write a proportion and solve it to answer this question.

Ratios can be used to express the grade of a hill or of a road. Figure 9-8 shows an example of this.

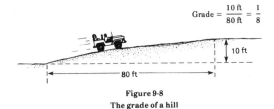

Figure 9-8
The grade of a hill

Applied Mathematics

- *Answer to the Study Activity:*
The solution to the tabletop problem is obtained as follows:

$$\frac{4"}{3"} = \frac{y"}{60"} : \text{all units are expressed in inches}$$

3y = 240, so y = 80 inches
Length of tabletop should be 6 ft 8 in.

Note: This problem could have been worked by using "inches to inches" for the ratio on the left, and "feet to feet" for the ratio on the right. But it seems safer, at this point in the students' development, to have them convert all numbers to the same unit.

- Answer to map problem:

$$\frac{1 \text{ inch}}{100 \text{ miles}} = \frac{4.25 \text{ inches}}{(y) \text{ miles}}$$

This gives y = 425 miles.

With regard to the same problem, you may want to point out that since one part of the scale is 1, we can simply measure on the model and multiply by the indicated ratio. In the problem posed for 1 inch represents 100 miles, the actual distance is 4.25 inches. Therefore, quite directly, 4.25 in. × 100 $^{mi}/_{in.}$ = 425 miles.

Using Ratios and Proportions
Page 21

A trail up the hill in Figure 9-8 would have a grade of 1 in 8, a very steep one! This way of expressing the rise or fall of land is also used with rivers. For example, the river that flows from Lake Okeechobee to the Florida Bay is about 60 miles long but only has a grade of 3 inches per mile, so that it flows very slowly and forms the great slowly-moving swamp called the Everglades. Since there are 63,360 inches in a mile, this is a grade of 3 in 63,360 or 1 in 21,120. A very gentle grade indeed!

Study Activity: In order for the flow to work properly in pipelines, the piping must have a certain grade or pitch. The pitch on one "run" of piping is such that the pipe drops 1 inch for every 4 feet. How many inches will this pipe drop in 92 feet?

The comparison of horizontal distance to vertical distance is used for the building of roofs as well as for hills, roads, rivers, and pipes. Carpenters use a comparison of vertical distance (rise) to horizontal distance (run) when they build roofs. Figure 9-9 shows this.

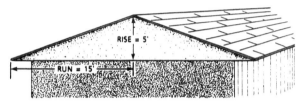

Figure 9-9
The slope of a roof

The slope of a roof is the ratio of the rise of the rafter to the run of the building.

$$\text{Slope} = \frac{\text{Rise}}{\text{Run}}$$

What is the slope of a roof that has a rise of 5 feet and a run of 15 feet?

Study Activity: You can find the height of a tree, flagpole, or building that is too high to measure easily by using a proportion. Figure 9-10 shows a picture of a person whose height you know standing next to a flagpole whose height you don't know. The ratio of "person shadow length" to "flagpole shadow length" is the same as the ratio of "person height" to

---

- *Answer to the Study Activity (pipeline problem):*

  The correct proportion is: $\frac{1 \text{ in.}}{4 \text{ ft}} = \frac{y \text{ in.}}{92 \text{ ft}}$

  Forming products of cross terms gives $4y = 92$, which when solved, yields $y = 23$. Since the proportion indicates that y is in inches, the answer tells us that the pipeline drops 23 inches in a 92-foot run.

- *Answer to question on slope of a roof.*
  For a roof with a rise of 5 feet and a run of 15 feet, the slope is:

  $$\frac{5 \text{ ft rise}}{15 \text{ ft run}} = \frac{1}{3} \text{ or } 1 : 3$$

- You may want to discuss proportions as they are used in scale models, scale drawings and maps. The scale may be a sentence like "1 inch = 50 miles." This is not good mathematical terminology. Point out that the expression means that 1 inch on the map **represents** 50 miles on land. Since 50 miles is equal to 3,168,000 inches, the actual ratio is:

  $$\frac{1 \text{ in.}}{3,168,000 \text{ in.}}$$

  Therefore, $\frac{\text{distance on the map}}{\text{distance on land}} = \frac{1 \text{ in.}}{3,168,000 \text{ in.}}$

"flagpole height." You can equate these two ratios to find the unknown height of the flagpole.

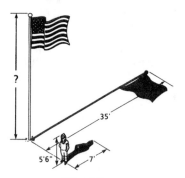

Figure 9-10
Using proportion to measure

If the person in the figure is 5 feet 6 inches tall, and casts a shadow that is 7 feet long, and the shadow of the flagpole measures 35 feet long, how tall is the flagpole?

Write the proportion in words first. Then substitute the numbers you know—remembering to change the 6 inches to feet. Then solve for the unknown term of the proportion. Is your answer a reasonable one? Does it have the correct units?

Two kinds of proportion

In the following list, as the quantity at the left increases, what happens to the quantity at the right?

miles driven in a car	time traveled
unit cost	total cost
distance on a map	actual distance
cups of flour used	cookies made
density	weight

If you answer that the quantity at the right **increases** as the one on the left **increases** you are correct. The quantities on the left and

---

- *Answer to the Study Activity (flagpole problem)*:
  The correct proportion in words and numbers is:

$$\frac{\text{length of flagpole's shadow}}{\text{length of person's shadow}} = \frac{\text{height of flagpole (y)}}{\text{height of person}}$$

$$\frac{35 \text{ ft}}{7 \text{ ft}} = \frac{y}{5.5 \text{ ft}}$$

Forming equal products of the cross terms:
$35 \times 5.5 = 7 \times y$, or $7y = 192.5$, giving $y = 27.5$ ft.
The flagpole is 27.5 feet high. The answer is reasonable and has the correct units—feet.

(An answer of 2.75 ft or 275 ft, would not have been reasonable.)

- It is <u>not</u> important that you stress the new vocabulary terms such as *pitch*, *rise*, *run*, *grade*, etc., or how to calculate them. Rather, simply introduce these terms as common words used in the appropriate trade area and define or explain them.

- You may need to use other examples if your students do not relate well to geometry, construction and carpentry. Applications that involve mixtures, food services and sampling may be helpful.

(Teacher's notes continued on page 28b.)

right are said to be **directly related**—if one increases, so does the other.

Now let's examine a special kind of direct relationship—called a **direct proportion**. Look at the formula given below.

$$\text{Distance} = \text{Rate} \times \text{Time}$$

or

$$D = R \times T$$

Suppose you travel at a constant rate of 50 miles per hour. How does *distance traveled* (D) depend on *time traveled* (T)? To help you answer this question, examine the following table.

R	T	D = R × T
50 mph	1 hr	50 mi
50 mph	2 hr	100 mi
50 mph	3 hr	150 mi
50 mph	4 hr	200 mi
50 mph	5 hr	250 mi

By looking at the table, you can see that in one hour, you travel 50 miles, in two hours you travel 100 miles, and so on. The distance traveled increases *in step*—or in a *fixed way*—with the time traveled. For example, if you double the time, you double the distance. If you triple the time, you triple the distance and so on.

Two quantities that increase together in step, or decrease together in step, are said to be directly proportional.

You can say, therefore, that *distance traveled* is directly proportional to *time traveled*. Can you also say that *distance traveled* for a given time, such as 2 hours, is directly proportional to the rate of travel?

Some quantities are related in a different way. In the following list, as the quantity at the left **increases**, what happens to the quantity at the right?

- We have taken some pains here to distinguish between a direct *relationship* and a direct *proportion*. They are not necessarily equivalent. For example, a direct relationship might involve a parabolic, hyperbolic or exponential relationship between two quantities, where as one increases, so does the other—*but not in step*. That is, one increases at a faster rate than does the other. In a direct proportion, the two *increase in step*, or *proportionally*, as described in the example involving the relationship D = R × T. Draw curves on the chalkboard—similar to the ones shown on the right—to make your point.

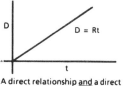

A direct relationship <u>and</u> a direct proportion between D and t.

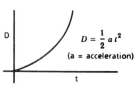

A direct relationship <u>but not</u> a direct proportion between D and t.

- It's worth some time going over the formula D = R × T and the accompanying table to show how the two quantities, D and T, are *proportionally* related, that is, they increase in step, and therefore are *directly proportional*.

  (*Teacher notes continued on page 28b.*)

Using Ratios and Proportions
Page 24

miles driven in a car	gasoline left in the tank
fuel burned	savings on the fuel bill
distance traveled	amount of tread on the tires
speed of a car	time needed to cover a distance

If you answer that the quantity at the right **decreases** as the one on the left **increases**, you are correct. The quantities on the left and right are said to be **inversely related**—if one increases, the other decreases.

Now let's examine a special kind of inverse relationship called an **inverse proportion**. Let's again use the formula that relates distance traveled, rate of travel and time of travel—but in a different form:

$$\text{Time of travel} = \frac{\text{Distance traveled}}{\text{Rate of travel}}$$

or

$$T = \frac{D}{R}$$

Answer this question—"How does time of travel (T) depend on the rate of travel (R) for any fixed distance?" To help you with the answer, examine the following table. Notice that the distance traveled is always constant—at 150 miles.

D	R	T = D/R
150 mi	10 mph	15 hr
150 mi	20 mph	7.5 hr
150 mi	30 mph	5 hr
150 mi	40 mph	3.75 hr
150 mi	50 mph	3 hr

By looking at the table, you can see that at 10 miles per hour, it takes 15 hours to go 150 miles, at 20 miles per hour it takes 7.5 hours, at 30 miles per hour it takes 5 hours, and so on. In other words, if you *double* the speed, it takes *one-half* as long, if you *triple* the speed, it

Applied Mathematics

- Go over the inverse proportions on this page. Show students that as the quantity specified in the left column increases, the corresponding quantity in the right column decreases—or is reduced.
- Just as you coupled the formula D = RT with its accompanying table to show a direct proportionality, couple the formula T = D/R and its accompanying table to show that for a fixed distance, the time T and the rate R are inversely proportional. Other inverse relationships in nature include:
  ▸ current (I) and resistance (R) in a circuit—at constant voltage (V): I = V/R
  ▸ acceleration (a) of mass (m) moved—at a constant applied force (F): a = F/m
  ▸ intensity of "point-source" light (I) at a receiver and distance squared (D^2) between point source and receiver (I = constant/D^2)
  ▸ force of attraction (F) between planets and distance squared (D^2) between the planets (F = constant/D^2)

Using Ratios and Proportions
Page 25

takes *one-third* as long, and so on. Increasing the speed of travel decreases the time of travel, *in step*, for any fixed distance.

Two quantities are said to be inversely proportional if one increases in step when the other decreases—or vice versa.

You can say, therefore, that for any fixed distance of travel the *time of travel* (T) is **inversely proportional** to the *speed of travel*. Can you write the formula $D = R \times T$ in another way to show another inverse proportion?

Example 2:
Solving an inverse proportion

Here is an example of a problem with an inverse proportion. If a car travels one way for 2 hours at 70 mph, how many hours will the return trip take if it travels at 30 mph? Although the quantities are inversely proportional, you work the problem in the same way you do for direct proportions, except now the pattern for the proportion is **reversed**.

Again, write the proportion in words first.

$$\frac{\text{Fast speed (trip 1)}}{\text{Slow speed (trip 2)}} = \frac{\text{Long time (trip 2)}}{\text{Short time (trip 1)}}$$

When quantities are directly proportional, the patterns of the two fractions always match. Here, because the quantities are inversely proportional, the pattern of the first ratio (trip 1 : trip 2) is reversed for the second ratio (trip 2 : trip 1).

Then substitute the numbers and find the unknown term.

$$\frac{70 \text{ mph}}{30 \text{ mph}} = \frac{y \text{ hours}}{2 \text{ hours}}$$

$$\frac{70}{30} = \frac{y}{2} \quad \text{(Units match so write only numbers.)}$$

$$30y = 140 \quad \text{(Equate products of cross terms.)}$$

$$\frac{\cancel{30}y}{\cancel{30}} = \frac{140}{30} \quad \text{(Divide each side by 30 to "free" y.)}$$

$$y = \frac{140}{30} = 4.67 \quad \text{(Use your calculator.)}$$

The return trip at 30 mph takes 4.67 hours, a much longer time. Check your answer. Is it reasonable?

Using Ratios and Proportions

- You can write the formula $D = R \times T$ in the form $R = D/T$, to show that for a fixed distance, the rate R is inversely proportional to the time T. Simply stated, this says that as the time required to travel a fixed distance *increases*, the rate (speed) required to travel the distance *decreases*, proportionally.

- Take some time to show students that the pattern is reversed in *inverse proportions*. Use chalkboard work with general inverse proportions like these,

$$\frac{(\ )_1}{(\ )_2} = \frac{(\ )_2}{(\ )_1} \quad \text{or} \quad \frac{(\ )_a}{(\ )_b} = \frac{(\ )_b}{(\ )_a}$$

stressing the flip-flop of 1,2 or a,b in the two equal ratios.

- In Example 2, the answer for y is $y = 4\,^2/_3$ hours, rounded to the decimal 4.67 hours. Students can show that the answer is both reasonable and accurate by testing the proportion

$$\frac{70}{30} = \frac{4.67}{2}$$

They can test it by calculating each ratio (using their calculators) and comparing results, or by forming and comparing the products of the cross terms.

Using Ratios and Proportions
Page 26

The use of a crowbar (lever) involves inverse proportions. The longer the lever arm of a crowbar, the less force it takes to do a job. Figure 9-11 shows the action of two crowbars of different lengths.

Figure 9-11
Crowbars illustrate inverse proportions

The crowbar with the **shorter** lever arm requires a larger applied force to remove the spike.

You can use proportions to work problems with crowbars such as those shown in the figure:

$$\frac{\text{Length of short bar}}{\text{Length of long bar}} = \frac{\text{Unknown force applied to long bar}}{\text{Force applied to short bar}}$$

Notice that since this is an **inverse** proportion, the *short bar* is on **top** of the fraction on the left but on the **bottom** of the fraction on the right.

Study Activity: If the lever arm of a short crowbar is 2 feet and 200 pounds of force are applied to it to remove the spike, how much force is required when a long crowbar whose lever arm is 3.5 feet is used? Check your answer for units and reasonableness, and make sure that the products of the cross terms are really equal.

The same principle applies to gears, as shown in Figure 9-12. The gear ratio is the ratio of the number of teeth on the drive gear to the number of teeth on the pinion gear.

26   Applied Mathematics

- In Figure 9-11, it is important to note that the distance (ℓ) from the fulcrum (point where the crowbar touches the floor) to the spike is *identical* in both drawings. That is, the crowbars differ only in length from the fulcrum to the "user" end, not from the fulcrum to the "application" end.
- Emphasize again the characteristic of inverse proportions by writing down

$$\frac{(\ )\,short\ bar}{(\ )\,long\ bar} = \frac{(\ )\,long\ bar}{(\ )\,short\ bar}$$

and showing the flip-flop of the words *short* and *long* in the two ratios.

- *Answer to the Study Activity (crowbar problem):*

$$\frac{200 \text{ pounds for short bar}}{y \text{ pounds for long bar}} = \frac{3.5 \text{ feet for long bar}}{2 \text{ feet for short bar}}$$

$$y = \frac{(2 \times 200)}{3.5} = 114.3 \text{ pounds}$$

The use of the long crowbar requires only 114.3 pounds to remove the spikes.
Checking the products of the cross terms gives:

$$\frac{200}{114.3} = \frac{3.5}{2}$$

(Teacher notes continue on page 28b.)

Using Ratios and Proportions
Page 27

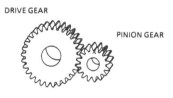

Figure 9-12
Gears illustrate inverse proportions

Study Activity: The angular speed of each moving gear is inversely proportional to the number of teeth on the gear. Suppose the drive gear has 50 teeth and the pinion gear has 20 teeth. If the drive gear makes 10 revolutions per minute, how many revolutions per minute will the pinion gear make?

Before you write a proportion, decide if the quantities in the problem are related directly or inversely. Direct proportions have the same comparison in the two ratios. Indirect proportions have the opposite comparison in the two ratios.

SUMMARY

When you compare two numbers or quantities, you are using *ratios*. These comparisons may be of a part to a whole, a part to a part or almost anything to another quantity.

Comparisons can be made with quantities that have the same or different units. When different units are used, the units must be included in the ratio.

When ratios involve time, they are often expressed with a unit of time in the denominator, such as feet per second or miles per hour. These special ratios are often called *rates*.

If ratios have the same kinds of measure (length, area, volume, etc.) you must be sure they are expressed in the same units of measure, like $^{3\,in.}/_{24\,in.}$ instead of $^{3\,in.}/_{2\,ft}$. Ratios that compare quantities with the same units are called *proper ratios*.

- *Answer to the Study Activity (gear problem):*

$$\frac{10 \text{ rpm (drive gear)}}{y \text{ rpm (pinion gear)}} = \frac{20 \text{ teeth (pinion gear)}}{50 \text{ teeth (drive gear)}}$$

$$20\,y = 50 \times 10 = 500$$

$$y = \frac{500}{20} = 25 \text{ rpm}$$

The pinion gear turns with an angular speed of 25 revolutions per minute. The fewer the teeth it has, the higher is its angular speed.

- After you have finished going over the direct-proportion and inverse-proportion variations, you might pose this problem. "How does a person's *body strength* vary with *age*?" This should result in a good class discussion. Ultimately, your students should decide that up to a certain age, body strength **increases** with age, and after that, body strength **decreases** with age. So this problem introduces them to the fact that two quantities need not **always** be related by purely a direct or inverse proportion. The proportion may switch from one to the other at a certain point. A graph of body strength versus age might look like this.

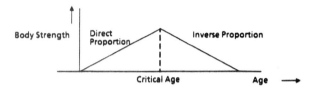

Using Ratios and Proportions
Page 28

Proportions are expressions that involve the equality of two ratios. They are often expressed as equivalent fractions, like $3/4 = 9/12$. Proportions are very useful to show relationships between corresponding parts of similar figures and between a scale drawing and the actual structure.

Proportions can be used to solve problems if only three of the terms in the two equal ratios are known. You learned to use a method of equating the products of cross terms in the proportion to find the missing term.

There are several important ideas for you to remember about proportions. Knowing these ideas will help you solve problems that involve proportions.

- Proportions show relationships between corresponding parts of similar figures.

- Proportions can be solved to calculate a missing or unknown term.

- Measures are **directly proportional** if one measure increases as the other increases, and decreases as the other decreases.

- Measures are **inversely proportional** if one measure increases as the other decreases, or decreases as the other increases.

Before you write proportions, decide if the quantities in the problem are related directly or inversely. Direct proportions have the same comparison in the two ratios. Indirect proportions have the opposite comparison in the two ratios.

Ratios show comparisons between quantities while proportions show equality between ratios. Proportions are very useful in building scale models, building designs, drawing and reading maps, and solving problems.

Applied Mathematics

- *Continued from page 27.*
- Again, by way of summary, be sure your students see the **order of terms** in a **direct** proportion:

$$\frac{(\text{Quant 1})_A}{(\text{Quant 1})_B} = \frac{(\text{Quant 2})_A}{(\text{Quant 2})_B} \text{ is } \frac{A}{B} = \frac{A}{B}$$

Point out that this order **reverses** on one side for an **inverse** proportion:

$$\frac{(\text{Quant 1})_A}{(\text{Quant 1})_B} = \frac{(\text{Quant 2})_B}{(\text{Quant 2})_A}$$

and becomes $\frac{A}{B} = \frac{B}{A}$

- The technique shown for setting up the direct and inverse proportions may make more sense to students if they think out what a "reasonable answer" should be. Some discussion with questions such as the following may be helpful:
  - Will a longer crowbar make it harder or easier to remove the spike?
  - Is there more or less gasoline used as the distance driven increases?
  - When we increase power in a car, what happens to the speed?
  - If more teeth are added to the pinion gear, does its angular speed increase or decrease?

(Teacher notes continued on page 28b.)

Teacher's Notes—Page 6 of Student Text
(continued)

- There are other constant ratios that you may want to discuss with your class. One that your students may understand is **density**, defined as mass per unit volume. Thus,

$$\text{density (water)} = \frac{1 \text{ gm}}{\text{cm}^3}$$

$$\text{density (lead)} = 11.34 \frac{\text{gm}}{\text{cm}^3}$$

$$\text{density (gold)} = 19.3 \frac{\text{gm}}{\text{cm}^3}$$

These ratios require units, else they are meaningless.
- *Proper ratios* often appear without units, such as 1 : 12 or $^1/_{12}$. The symbol π is a number that stands for the proper ratio $^c/_d$.

Teacher's Notes—Page 7 of Student Text
(continued)

- Time-related ratios are an important concept for your students to understand. Help them gain this understanding by emphasizing examples like these:
  ▸ *Linear speed*: ft/sec, meters/sec, miles/hr, etc.
  ▸ *Fluid flow*: gal/min, ft³/sec; liters/min
  ▸ *Heat energy flow*: calories/sec, Btu/hr
  ▸ *Rates of production*: candy bars per hour, autos per day, tons of coal per day.
  ▸ *Rates of increase or decrease*: dollars lost per year, bushels of grain per year.
  ▸ *Angular speed*: revolutions per minute, radians per second.
- *Answer to the Study Activity*:
  The example of a bicycle messenger averaging 10 miles in two hours would have a "distance-to-time ratio" of 10 miles/2 hours or 5 mph. The computation is trivial, but stress the fact that 5 mph is a better way to report the rate than 10 miles/2 hours. Tell your students that rates are usually reported with a single unit of time—per minute, per second, per hour—in the denominator.

Teacher's Notes—Page 12 of Student Text
(continued)

For example, sides s and q in triangle A correspond respectively to sides g and f in triangle B—and the angles between sides s and q, and between g and f, are identical.

- You may want also to stress the following ideas as you discuss **similarity** with your students:
  ▸ All circles are similar.
  ▸ Similar polygons (straight-line, closed figures) have corresponding equal angles.
  ▸ If the corresponding sides of 2-dimensional figures are proportional, their areas are proportional.
  ▸ If the corresponding sides of 3-dimensional figures are proportional, their volumes are proportional.
  ▸ Similar figures are scale models of each other (scaled up or scaled down).

Teacher's Notes—Page 15 of Student Text
(continued)

For example, ask students to examine the following expressions and identify which are valid proportions.

1. $\dfrac{a}{b} = \dfrac{d}{e}$ 2. $\dfrac{a}{b} = \dfrac{d}{f}$

3. $\dfrac{c}{f} = \dfrac{b}{e}$ 4. $\dfrac{a}{c} = \dfrac{f}{e}$

5. $\dfrac{b}{d} = \dfrac{a}{f}$ 6. $\dfrac{b}{e} = \dfrac{a}{d}$

7. $\dfrac{e}{b} = \dfrac{f}{c}$ 8. $\dfrac{a}{d} = \dfrac{c}{f}$

(1, 3, 6, 7 and 8 are valid proportions).

Teacher's Notes—Page 16 of Student Text
(continued)

With b known, any of the following proportions will determine a:

$$\frac{3}{5} = \frac{9}{a} \text{ ; } \frac{4}{5} = \frac{12}{a} \text{ ; } \frac{12}{4} = \frac{a}{5} \text{ ;}$$

$$\frac{a}{5} = \frac{12}{4} \text{ ; } \frac{a}{9} = \frac{5}{3} \text{ ; etc. ; } a = 15$$

- With the introduction of y for the unknown in the proportion, we are beginning to use algebraic notation. Take a few minutes to explain that writing "y" is no different from writing the symbol "?", and writing letters—not punctuation marks—is the accepted practice.

Using Ratios and Proportions

Teacher's Notes—Page 22 of Student Text
(continued)

Read through the 40 student exercises to get ideas for ratios and proportions used in the areas of agriculture, business and marketing, home economics and health occupations.

- The proportions discussed up to this point have been mostly part-to-part and part-to-whole. With these kinds of proportions, the comparisons show scaled-up or scaled-down situations. These kinds of relationships are **direct** relationships. If one part of a similar figure is larger than the part of another figure it's compared to, then all parts of that figure are correspondingly larger than the parts of the other figure. If one part of a similar figure is smaller, then all parts of that figure are correspondingly smaller.
- Go over the list of direct proportions and make sure your students "see" the direct proportion in each one. They should understand that as the quantity specified in the left column increases, the corresponding quantity in the right column increases also.

Teacher's Notes—Page 23 of Student Text
(continued)

- The table shows that *distance traveled* is *directly proportional* to *time involved*, for a given rate. It is also true—or can be shown in another table—that for a given time, *distance traveled* is *directly proportional* to the *rate of travel*.

Teacher's Notes—Page 26 of Student Text
(continued)

$200 \times 2 = 114.3 \times 3.5$ (Remember, 114.3 is a rounded number!)
$400 = 400.05$ (near enough to show equality)

- Go over the **inverse relationship** between applied force and lever-arm length in the crowbar use—as illustrated in Figure 9-11. Then, to test your students' understanding, ask them why the doorknob on a door is placed near the **vertical edge** of the door **opposite** the hinges—and not, say, at the center, a position nearer to the hinges. If they understand the idea of reduction in applied force with increasing lever arm, they should point out that it's to increase the lever arm and therefore to reduce the force needed to open a door.

Teacher's Notes—Page 28 of Student Text
(continued)

These kinds of questions will increase their understanding of *direct* and *inverse* relationships.

- As a final note, you might want to consider the following: Point out to your students that care is to be exercised in measuring distances on a scale model or land map when **scaling up** is involved. That's because one multiplies the measurement made on the model or map by the scale factor to get the scaled-up dimension. Thus, any error in the measure of the scaled model is **amplified** by the scale factor.

The following example will illustrate this:

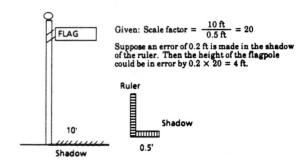

Given: Scale factor = $\frac{10 \text{ ft}}{0.5 \text{ ft}} = 20$

Suppose an error of 0.2 ft is made in the shadow of the ruler. Then the height of the flagpole could be in error by $0.2 \times 20 = 4$ ft.

PRACTICING THE SKILLS

Laboratory Activities
Use the mathematics skills you have learned to complete one or all of the following activities:

Activity 1: Ratios and proportions in similar triangles

Equipment Measuring tape
String
Masking tape
Carpenter's square
Two ¼-inch dowel rods of lengths 18 in. and 36 in.
Calculator

Statement of Problem In this activity, you compare the measurements of corresponding sides of similar triangles.

Procedure First, you lay out a large triangle in your classroom, using a wall, floor and piece of string to form the three sides. Then you position a dowel rod within the large triangle to form a smaller triangle that is similar to the large triangle. (See drawing.)

a. <u>Laying out the large triangle.</u> Measure a 10-ft section of string. Measure a distance 6 feet straight up on a classroom wall. Mark this position (A) with a piece of masking tape. (See drawing.) Have a member of the group hold one end of the 10-ft string on the wall at the marked position. Pull the string tight and move the other end until it touches the floor. Use a piece of masking tape to mark the position where the end of the string (C) touches the floor. Measure the distance along the floor, from B directly beneath the end of the string, to the end of the string touching the floor at C. What is the distance BC? What special type of triangle is the large triangle you have made? Record your answers on a separate sheet of paper.

Using Ratios and Proportions

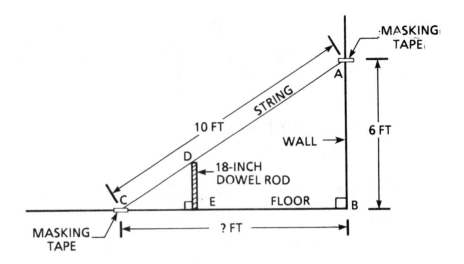

b. <u>Setting up similar triangles.</u> Measure the length of each dowel rod and write the measurements on a sheet of paper. Choose the shortest dowel rod and move it along the floor, underneath the string, until the top of the rod just touches the string at D. Use the carpenter's square to be sure the dowel rod is perpendicular to the floor. If the rod is not perpendicular to the floor, adjust its position until it is—with the rod top just touching the string. With masking tape, mark the bottom of the dowel rod on the floor at E and the top of the rod on the string at D. Measure the distance (DC), along the string from the masking tape to the end of the string on the floor. Record this measurement. Measure the distance (EC) along the floor, from the bottom of the rod to the masking tape. Round this measurement.

c. Repeat Step b for the other dowel rod.

Calculations

a. Write ratios for the **length of the rod** (DE) to the **length of the string segment** (DC) for each set of measurements.

b. Are the ratios equal?

c. Write ratios for the **length of the rod** (DE) to the **distance on the floor** (CE) for each set of measurements.

d. Are the ratios equal?

e. Write ratios for the **length of the string segment** (DC) to the **distance on the floor** (CE) for each set of measurements.

f. Are the ratios equal?

g. Compare the ratios calculated in a, c, and e to the **corresponding ratios** for the large triangle formed by the wall, the string, and the floor. Each of these equal ratios forms **proportions**. Can you write other proportions using these measurements?

Activity 2:	Using ratios to make scale drawings
Equipment	Drawing kit (Accu-Line™) Measuring tape Paper and pencil Ruler Calculator
Statement of Problem	In this activity, you use proportions to make a scale drawing of the classroom floor and of one of the classroom walls.
Procedure	**a.** Measure the length and width of the classroom. Write these measurements on a sheet of data paper.
	b. Measure length and width of the teacher's desk and the distance from one corner of the desk to each of the two nearest walls. Write these measurements on a sheet of data paper.
	c. Choose a classroom wall with several windows—or a wall with a door if no windows exist. Measure the length and height of the wall. Measure the length and height of one window or door. Make appropriate measurements to locate the position of the window or door in the wall. Write these measurements on a sheet of data paper.
Calculations	**a.** Use the drawing kit (Accu-Line™) to draw the floor plan of the classroom, showing the position of the teacher's desk. To draw the floor plan, use the floor measurements you made and a scale of *1 inch to 4 feet*.
	b. Using the same scale and the wall/window measurements you made, draw the wall plan accurately, locating the windows or door where they belong.

Activity 3: Lever arm and force ratios

Equipment
Meter stick
Two spring scales with 5000-gram capacity
Tape measure
Fulcrum (a wooden block with a triangular cross section and a height of about 4 inches).
Calculator

Statement of Problem
In this activity, you determine ratios of input and output forces using a lever. The ratios are found for three positions of the fulcrum and tested to determine if they are proportional.

Procedure
a. Place the 40-centimeter mark of the meter stick on the fulcrum and position the fulcrum and meter stick so that the two ends extend off the table or desk, as shown in the drawing below. Hook one spring scale as close to the 0-cm mark as possible (this is the output force). Hook the other spring scale—for the input force—as close to the 100-cm mark as possible. (You may have to use some masking tape to be sure the hooks don't fall off the meter stick.)

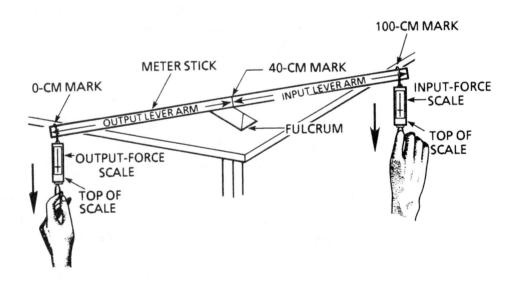

b. Hold the top of the *output-force* scale near the edge of the desk as shown. Pull down on the *input-force* scale until you have an input force reading of 1500 g. Read the output-force scale reading. Write these values on a sheet of data paper. Repeat this process for input-force readings of 2500 g and 3000 g.

c. Move the fulcrum to the 60-cm mark on the meter stick and repeat Part b.

d. Move the fulcrum to the 80-cm mark on the meter stick and repeat Part b.

Calculations

a. Write the ratio of input force to output force for each fulcrum position. Are these ratios for a given fulcrum position equal? Do they form proportions?

b. Are the ratios at the different fulcrum positions of 40 cm, 60 cm and 80 cm equal to each other? Do they form proportions?

c. In this setup the output force has been located at the 0-cm position on the meter stick. Therefore the length of the output lever arm—from output force to fulcrum—is given by the meter stick reading at the fulcrum position. The input force is located at the 100-cm mark. Therefore the length of the input lever arm—from fulcrum to input force—is equal to 100 cm minus the reading at the fulcrum position. Find a relationship that uses the input and output *forces* and the input and output *lever arms* to form a proportion. Is the proportion a *direct* or *inverse* proportion?

Student Exercises

You can solve the exercises that follow by applying the mathematics skills you've learned. The problems described here are those you may meet in the world of work.

NOTE: *Wherever possible, use your calculator to solve the problems that require numerical answers.*

 GENERAL

EXERCISE 1: A common proportion is the percentage, where a quantity is compared to 100. Suppose on your last paycheck, your gross pay was $427.50, and $51.01 was deducted for medical insurance.

a. What is the ratio of medical insurance deduction to your gross pay?

b. Equate this ratio to a second ratio having 100 as the denominator. What must the numerator in this second ratio be for the two ratios to be equal?

c. What percentage is being deducted for medical insurance?

EXERCISE 2: A nut mix is to be prepared with cashews, almonds, and peanuts in the ratio of 1 : 1 : 3 by weight. If you have 8 ounces of cashews, how many ounces of almonds and peanuts are needed? What will be the total weight of this mix?

EXERCISE 3: A state's automobile inspection program checks headlight alignment. The drop in the light beam cannot be greater than 2 inches for each 25 feet in front of the car. Suppose the headlights on your car are 28 inches above the ground. See the illustration below.

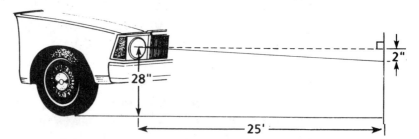

a. What is the ratio of drop of the light beam to the distance in front of the car permitted by the inspection program?

b. Suppose your headlights were adjusted to this ratio. How far in front of your car would the light beam from your headlights strike the ground?

EXERCISE 4: For each of the pairs below, state 1) whether the relationships are normally found to be **directly**, or **indirectly related**, and 2) whether or not the two ratios are **proportional**. (Round final results to the nearest whole number when testing proportionality.)

  a. Gross pay of $420.00 vs FICA withholding of $30.03
  Gross pay of $617.20 vs FICA withholding of $44.13

  b. Air conditioner thermostat setting on 68°F vs $103.12 electric bill
  Air conditioner thermostat setting on 78°F vs $88.54 electric bill

  c. 30' height of flagpole vs 26' length of shadow
  6' height of person next to flagpole vs 5.2' length of shadow

  d. 12-oz can soft drink vs $0.50 cost
  32-oz bottle of soft drink vs $0.89 cost

  e. $4.80 interest paid to savings account vs $480.00 average balance
  $11.70 interest paid to savings account vs $936.00 average balance

EXERCISE 5: A computer desk hutch (shelves that sit atop a desk) is labeled as being 38 ¾ inches tall. The illustration on the box shows a fixed shelf a little more than halfway up. Your computer is 21 ½ inches tall. Before you purchase the hutch, you need to know if your computer will fit under the fixed shelf. Since the height of the fixed shelf is not provided on the box, you decide to estimate the height by measuring the illustration. In the picture on the box, the hutch appears 23.2 cm tall, and the space beneath the fixed shelf is 11.4 cm.

  a. Would you expect the relationship between the dimensions in the picture and the real dimensions to be directly related or indirectly related?

  b. Write in words the ratio of the illustration's dimensions that involves the shelf height and the overall height of the hutch. Write a similar ratio in words for the real dimensions of the shelf height and hutch.

c. Substitute the values you know for the words in the ratios. Since you expect the sizes in the picture to be proportional to the real sizes, set up a proportion and find the real dimensions for the shelf height. Is there enough room for your computer to fit beneath the fixed shelf?

AGRICULTURE AND AGRIBUSINESS

EXERCISE 6: For good maintenance of lawns, a fertilizer analysis ratio of 3 : 1 : 2 is commonly used. This is the ratio of the fertilizer's content of nitrogen to phosphate to potassium. The strengths of each component are reported similarly on fertilizer bags in terms of percentages by weight. Several different mixes are available: 28-4-4, 21-0-0, 15-5-14, 12-4-8, 13-13-13, and 12-24-12. Which fertilizer mix (or mixes) has the desired ratio of 3 : 1 : 2?

EXERCISE 7: Several ratios have been developed to analyze the financial strength and growth of farm businesses. One of these is the current ratio, a ratio that indicates the extent to which current liabilities could be liquidated, if necessary. Most businesses are expected to maintain a current ratio of at least 2 to 1.

$$\text{Current ratio} = \frac{\text{Total current assets}}{\text{Total current liabilities}}$$

a. An analysis of the Browns' farm business shows their current assets total $47,206. Their total liabilities are $9498. Compute the Browns' current ratio.

b. The previous year's balance sheet for the Browns showed assets totaling $38,314 and liabilities totaling $7908. Compute the current ratio for the previous year.

c. Has the Browns' financial position strengthened this year (a higher current ratio) or weakened (a lower current ratio) when compared to the previous year?

EXERCISE 8: A tractor has a belt pulley diameter of 10 inches that operates at a speed of 1100 revolutions per minute (rpm). The pulley is to be connected to a forage blower that needs to operate at a speed of about 650 rpm. What size pulley should be used on the forage blower?

EXERCISE 9: The container of lawn herbicide (weed killer) instructs you to prepare a solution of 1 part full-strength herbicide to 10 parts water when using a hose-end sprayer.

a. What is the ratio of full-strength herbicide to water?

b. Your hose-end sprayer holds up to 16 oz of mix. If you fill the sprayer with 14 oz of water, how many ounces of full-strength herbicide should you add to obtain the proper concentration?

c. Suppose for direct application, the instructions are to mix 1 oz of full-strength herbicide to each gallon of water. What is the ratio of full-strength herbicide to water in this instance?

d. How does this compare to the strength of solution prepared in the hose-end sprayer?

EXERCISE 10: The table below shows that hay may be substituted for the more expensive grain, and still produce a similar effect on a feeder steer. Notice that for each change of 100 pounds of hay in the first column, the amount of grain needed to create the same effect is reduced.

Combinations of grain and alfalfa hay to produce 300-lb gain on a 700-lb steer

Pounds hay	Pounds grain
1000	1316
1100	1259
1200	1208
1300	1162
1400	1120
1500	1081
1600	1046
1700	1014
1800	984
1900	957

a. Copy the table onto your paper. Beginning with the second line of the table, compute the amount of grain that has been substituted for 100 pounds of hay in each line of the table. Record this in a third column on your table labeled Change in grain.

b. For each line of your table, compute the ratio of change in grain to the change in hay. Add this to a fourth column in your table. Label the column for the name of this ratio the Marginal rate of substitution.

c. The price ratio (in this case) is defined as the ratio of the price of hay to the price of grain. Suppose grain costs you 15¢ per pound and hay costs you 6¢ per pound. What is your price ratio?

d. To have the lowest cost combination, the marginal rate of substitution must equal or be greater than the price ratio. Which substitution of hay for grain (from your table) should you select as the lowest cost combination?

BUSINESS AND MARKETING

EXERCISE 11: A newspaper photographer took a picture of the new statue in front of city hall. The statue is 4 ½ feet wide and 17 feet tall. The copy editor has room for a photograph that is 4 inches wide.

a. What is the ratio of the statue's width to its height?

b. If the size of the photograph has this same ratio of width to height, how "tall" will the photograph be that fits into the width allowed by the copy editor?

EXERCISE 12: One measure of the ability of a business to meet its current debts is the *current ratio,* the ratio of current assets to current liabilities, as shown below.

$$\text{Current ratio} = \frac{\text{Total current assets}}{\text{Total current liabilities}}$$

Most businesses are expected to maintain a current ratio of at least 2 to 1. The balance sheet of your air conditioner repair service shows your business' current assets total $32,625, and your current liabilities total $19,480.

a. Calculate your current ratio.

b. Based on your current ratio, how would a bank feel about loaning money to your business for expansion?

EXERCISE 13: Inventory turnover is the ratio of the cost of goods sold to the average inventory value, as shown below.

$$\text{Inventory turnover} = \frac{\text{Cost of goods sold}}{\text{Average inventory value}}$$

A ratio of 1 : 1 means that during the period in question the business sells an amount equivalent to the amount it stores in inventory. A smaller ratio suggests too much inventory that is not being sold. A higher value suggests a high turnover in the inventory—usually desirable for a business.

Glenn's Office Supplies started the year with an inventory worth $28,437, and ended the year with an inventory worth $33,010. During the year the business sold $263,841 worth of office supplies.

a. Compute the average inventory value for the year, using the starting and ending values given above.

b. What is the inventory turnover ratio for Glenn's Office Supplies? (Round to 1 decimal place.)

c. Interpret the inventory turnover ratio you obtained. Would you say that the store is a "low inventory, fast turnover," or a "high inventory, slow turnover" business?

EXERCISE 14: West and Brown are partners. West invested $22,000 and Brown invested $48,000 in the business. They agree to split the first-year profit of $89,600 in proportion to their initial investments.

a. What is the ratio of each partner's contribution to the total initial investment?

b. What portion of the total first-year profit should each partner receive?

EXERCISE 15: The standard aspect ratio (width to height) for a television picture is 4 : 3. The program director has scheduled to broadcast a wide-screen movie that was filmed with an aspect ratio of 2 : 1.

a. Draw a sketch of a rectangular-shaped television screen on your paper that has the standard aspect ratio for a television broadcast.

b. Suppose the wide-screen movie is to be broadcast so that the picture occupies the full television screen height. Draw a second rectangle on top of the first that has a 2 : 1 aspect ratio and has the same height.

c. Use your drawing to compute what portion of the wide screen movie picture's width will be lost at the sides. Express this portion as a percent.

d. Suppose that instead of showing the full height of the movie, you wished to show the full width but maintain the correct aspect ratio of 2 : 1. What would be the result on the television screen?

EXERCISE 16: A typing pool has been increasing the number of employees and word-processing terminals. Susan keeps the time-sheet records for the typists, and prepares a summary of the number of typists employed and the hours logged on each terminal per week, as shown below.

Weekly terminal usage in the typing pool

Number of typists	Hours terminal usage
6	230
7	253
8	272
9	285
10	310

a. Write the ratio of hours the terminals are used to the number of typists employed for each entry in the table. Compute the decimal fraction for each ratio.

b. What trend is apparent from the changing ratios?

c. Do these changing ratios indicate a direct or indirect relationship between the number of typists and the hours of terminal usage?

EXERCISE 17: As manager of a small print shop, you deal with many types of paper. You usually use international (ISO) paper sizes. Most common is the A-series used for general printed matter such as publications and stationery. The illustration below shows the relationships between the various A-series sizes, A0 through A10. The A0 size is specified to have an area of 1 m^2, and measures 841 mm wide and 1189 mm long. Each successive size is half the area of the preceding size. You can obtain each size by folding the preceding larger size in half.

a. What is the ratio of the width to the length, based on the dimensions of the A0 paper?

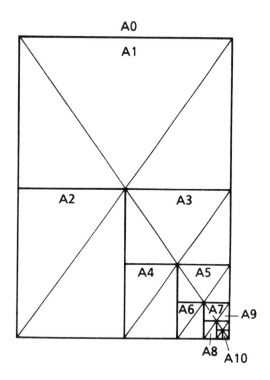

b. If you were to make a single cut in the A0 paper, you would obtain two pieces of A1 paper. What would be the ratio of width to length of the A1 paper?

c. Suppose you had envelopes designed to hold A2 paper. How could you fold a printed greeting card made with A0 paper to fit in the envelope? What do you think the ratio of width to length will be for the greeting card?

EXERCISE 18: You have a total of 68 cars on the lot for sale. A breakdown of your inventory is shown in the table below.

Inventory of cars on lot

Model	Inventory
Selvo	12
Thistle	28
Cascade	20
Luxurie	8

Your home office recommends that its franchises should stock the inventories of the Selvo, Thistle, Cascade, and Luxurie models in a ratio of 3 : 7 : 5 : 2. Is the inventory on your lot in agreement with the home office recommendation?

Using Ratios and Proportions

EXERCISE 19: You are named as executor of a friend's estate. Your friend has three children, and in his will he asks that his estate be divided among his three children in proportion to their ages at the time of his death. The amount of his estate is $107,065, and the ages of the three heirs are 52, 48, and 33 years old.

    a. Write the ratio of the age of each heir to the *total* of all their ages. According to the will, each of these ratios will be equal to a second ratio: the ratio of the amount inherited to the total estate value.

    b. Determine the amount each heir will inherit by solving the proportions made with the ratios of Part a.

    c. How can you check your calculation of each portion of the inheritance?

EXERCISE 20: You manage a mail-order business that has 5 workers assigned to process and package the orders it receives. These workers are packaging an order for 2500 parts that need to be mailed by the end of the shift. So far they have packaged 900 parts during the first four hours. It is evident that they are not going to be able to finish the order during the remaining four hours without some help.

    a. What is the ratio of parts packaged to the number of workers assigned during the first four hours? What is the meaning of this ratio when expressed in simplest form?

    b. How many parts still have to be packaged during the remaining four hours?

    c. How many total workers will it take to be able to meet this goal? (Assume all workers will work at the same rate computed above.)

    d. How many additional workers will you need to reassign during the last half of the shift?

HEALTH OCCUPATIONS

EXERCISE 21: A particular saline solution is prepared by mixing 1 part salt in 25 parts water.

    a. What is the ratio of salt to water in such a solution?

    b. If you mixed 1 part salt with 25 parts water, how many "parts" total would you have? How many parts of salt are there in 100 parts of this solution?

    c. What percent salt is such a solution?

EXERCISE 22: Red blood cells should account for 99.98% of all blood cells, with 0.02% accounting for all other cells.

    a. What is the ratio of red cells to all other cells?

    b. A blood analysis showed 4,600,000 red cells and 6000 other cells. Is this blood cell count close to normal? If not, is the other cell count higher or lower than normal?

EXERCISE 23: During the first year of operation a clinic treated a total of 4836 patients, and gave 622 flu immunizations. This year the clinic is seeing many more patients. During the first six months the clinic has treated 3418, and given 481 flu immunizations.

    a. What is the ratio of flu shots to patients seen during the first year of the clinic's operation? during the first six months of the second year?

    b. Regarding the amount of flu shots now being given, is the clinic giving proportionally the same number of shots, more shots, or less shots than the first year it was open?

EXERCISE 24: An ambulance service is analyzing the need for additional vehicles. One of the ambulance crews has found that it can reach the scene of an accident faster at night than in the daytime. They maintain that another unit might be needed during the daylight hours. To prove this the crew shows the following data to their supervisor.

Times to reach scene of accident

	Distance to accident (mi)	Time to arrive (min)
Daytime calls	2.7	5.5
	6.2	11.1
	3.9	8.8
Nighttime calls	4.0	6.2
	8.3	10.0
	2.1	3.3

a. Compute the rate of travel (the ratio of distance traveled to time needed to travel the distance) for each of the calls.

b. Does the claim of the ambulance crew seem justified by their data?

c. Obtain an average rate for the daytime calls. Do the same for the nighttime calls. What is the ratio of the average nighttime rate to the average daytime rate?

d. Complete this sentence (rounded to one decimal place):
When answering nighttime calls, the crew can drive to the scene about _____ times faster than on daytime calls.

EXERCISE 25: A water treatment plant deals with very large volumes of water, and very small concentrations of chemical agents. Under normal conditions, a particular chemical is added to the water to obtain a concentration of 0.9 parts per million. The treatment facility has holding tanks that contain 700,000 gallons.

a. What is the desired ratio of gallons of chemical to gallons of water?

b. How many gallons of this chemical should be added to the holding tank to obtain the desired concentration?

HOME ECONOMICS

EXERCISE 26: You are catering a wedding and deciding how much punch will be needed. Normally for a reception with 75 guests you prepare 5 gallons. This reception however is much bigger. You are expecting almost 300 guests. About how much punch should you plan to prepare?

EXERCISE 27: A standardized clothing size system has been proposed, based on metric measurements. A part of the proposal is to standardize men's shirt sizes based on the neck measurements in centimeters, as shown below.

Proposed men's shirt sizes

Size	Neck (cm)	Chest (cm)
37	37	86
38	38	90
39	39	94
40	40	98
41	41	102
42	42	106
43	43	110
44	44	114
45	45	118
46	46	122

a. The chest size appears to increase as the neck size increases. What is the ratio of the neck size to the chest size for the smallest shirt size shown? the largest shirt size?

b. Does the ratio stay the same for all the sizes?

EXERCISE 28: Many of the calories found in foods come from the fat content. Each gram of fat contributes about 9 calories. Analyzing the proportion of calories in foods that come from the fat content can be an aid in choosing wholesome foods.

a. The label on a box of cereal reports that a one-ounce serving contains 130 calories and 6 g of fat. Compute the number of calories contributed by the fat and the ratio of fat-contributed calories to total calories.

b. Another cereal brand advertises 110 calories per serving and 4 g of fat. Compute the number of calories contributed by the fat and the ratio of fat-contributed calories to total calories.

c. Which cereal has proportionally more of the calories coming from the fat content of the cereal?

Using Ratios and Proportions

EXERCISE 29: When caring for indoor plants, the proper mix of plant food is important. The label on a liquid concentrate instructs you to mix 5 ounces of concentrate in a gallon of water.

a. What is the ratio (with the same units, or no units) of concentrate to water?

b. Suppose you only wanted to prepare about a quart of fertilizer solution. How much concentrate should you add to a quart of water?

EXERCISE 30: A family has two children, ages 7 years and 2 years. In preparing for the children's college education, the parents begin savings accounts for each. However, since the older will be ready to attend college 5 years sooner than the younger, the parents feel that they should contribute a larger portion to the older child's account each month. They have a total of $100 per month to contribute to the children's accounts. They propose to split the money proportionally, based on the children's ages.

a. Compute the ratio of each child's age to the total of their ages during the first year of the plan. How much of the $100 will each child's account receive each month during this first year?

b. Compute the ratio again for the twelfth year of the plan, when the older is 18 years old. How much of the $100 will each child's account receive during this twelfth year?

c. Does this seem fair? What should the parents do for the younger child's account after the older reaches 18?

INDUSTRIAL TECHNOLOGY

EXERCISE 31: A 2"-diameter pulley is mounted on a motor shaft that turns at 1075 revolutions per minute (rpm). This pulley is connected by a belt to another pulley mounted to the shaft of a fan blade assembly. The fan pulley is 5" in diameter. Pulley diameters and speeds are inversely proportional, like gears. How many revolutions will the fan blade make each minute?

EXERCISE 32: Gears are frequently described with ratios. For example, a certain car's transmission and differential has a total gear reduction of 3.4 : 1. That is, the ratio of the crankshaft speed to the drive axle speed is 3.4 to 1. If the crankshaft in this car is turning at 1500 revolutions per minute (rpm), how fast would the drive axle be turning?

EXERCISE 33: A cutting oil is prepared with two oils from the stockroom, Code 171 and Code 209, in the ratio of 8 to 15.

a. If you started with 16 oz of Code 171, how much of Code 209 should you add?

b. How many ounces total would your batch be?

c. What fraction of the total batch is Code 171? Code 209? (Express your answers as percentages.)

EXERCISE 34: An electrical transformer can be constructed by wrapping two or more wires around an iron core. An input voltage is applied to the pair of wires called the "primary coil" and an output voltage is obtained from the pair of wires called the "secondary coil." In a certain transformer, the primary coil makes 10,000 turns around the iron core. The secondary coil make 1000 turns. A voltage of 120 volts (ac) is applied to the primary coil and you observe an output voltage from the secondary coil of 12 volts.

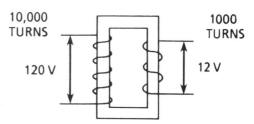

a. Write the relationship between the voltage and the number of turns of wire in each coil as a proportion.

b. Are the primary and secondary voltages directly or indirectly proportional to the number of turns of wire?

c. What output voltage should you observe if you reversed the setup, and applied 120 volts to the coil with 1000 turns?

EXERCISE 35: A mason mixes mortar in the ratio of 1 part cement to 3 parts sand (by volume). The mixer tub can hold 0.65 m³. Each bag of cement measures about 0.05 m³.

a. How much sand should the bricklayer add with each bag of cement to obtain the proper mix for the mortar?

b. How many whole batches can be mixed at one time in the mixer? (One "batch" uses one bag of cement.)

EXERCISE 36: Blueprints and floorplans of buildings are not normally drawn actual size. Instead they are "scaled down," so that a 1 foot distance in the building may appear as a ¼" distance on the drawing, for instance.

a. What is the ratio that describes the scale used in such a drawing? (Show the ratio both with and without dimensions.)

b. How long a line on the paper would be used to represent a wall that is 24' long?

c. How long is a duct that is depicted on the drawing by a 9 ½-inch line?

EXERCISE 37: A catalog lists the characteristics of several manually operated winches. One of the features listed is the gear ratio. The worm gear shown below is advertised to have a gear ratio of 41 : 1. That is, 41 turns of the hand crank are needed to produce one turn of the large gear.

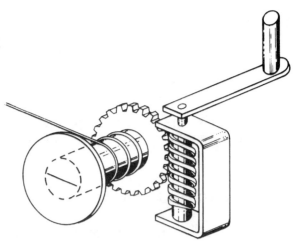

a. The drum attached to the large gear has a diameter of 1 ½" and is used to reel in a length of cable. About how many times would the hand crank have to be turned to reel in 12" of cable? (Round to the nearest whole turn of the crank.)

b. Suppose you are able to turn the hand crank at a rate of 40 turns per minute. About how many minutes would it take you to wind 12" of cable? (Round to the nearest 0.1 minute.)

EXERCISE 38: A "pressure intensifier" is used to boost pressures. As shown below, two pistons are attached with a push rod. The pressure P_i in the left chamber pushes on the piston that has an area A_i. The push rod presses on the second piston having an area A_o and creates a higher pressure P_o. **The ratio of the pressure on the two pistons is indirectly (or inversely) proportional to the ratio of the areas of the pistons.**

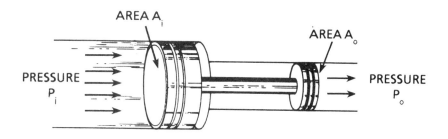

a. Suppose the area of the left piston surface (A_i) is 25 in² and the area of the right piston (A_o) is 4 in². If the input pressure (P_i) is 35 pounds per square inch (psi), what will be the output pressure (P_o)?

b. What is the result on the output pressure (P_o) if the input pressure (P_i) is doubled?

c. What is the result on the output pressure if the area of the output piston (A_o) is doubled to 8 in²?

EXERCISE 39: The steepness of a railroad track over a 3-mile distance is reported as a rising grade of 1 in 42. This means it rises 1 foot for every 42 feet of track. It is followed by another 5 miles of track that has a descending grade of 1 in 78.

a. Which of the two distances has the steeper grade?

b. How much altitude does the train gain when traveling over the first 3-mile distance? (Express your answer in feet.)

c. Does the train lose all that altitude after traveling back down the following 5 miles of track?

EXERCISE 40: The compression stroke of an engine's piston is often described by the ratio of the volume before the compression begins to the volume at the end of the compression, for example "10 to 1," or "17 to 1." This is known as the compression ratio.

a. Suppose a cylinder at the beginning of the compression stroke had a volume of 8 cubic inches, and was compressed to a volume of 1 cubic inch. What is the compression ratio of this cylinder?

b. Suppose you measure the cylinder of a large diesel engine. The cylinder has a diameter of 6.250 inches and a stroke of 8.000 inches, as shown below. At the end of the compression stroke, there remains a small 15.970-in³ volume. Compute the volume of the diesel-air mixture at the start of the compression stroke, and the resulting compression ratio for this cylinder. (Reduce the ratio to a form with "1" as the denominator, similar to the examples above.)

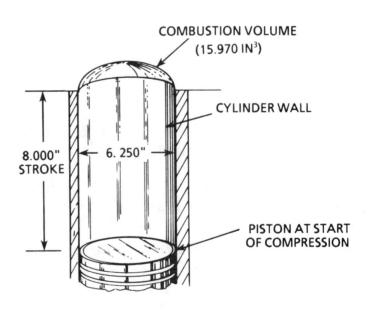

GLOSSARY

Constant ratio A constant ratio is a ratio like $^{circumference}/_{diameter}$ that has the same value (π) in all cases. The density of elements—such as 11.34 $^{gm}/_{cm^3}$ for lead—and the speed of light, such as 186,000 $^{miles}/_{sec}$—are examples of constant ratios.

Direct proportional relationship A relationship between variables like voltage, current and resistance where $V = I \times R$. The current (I) increases as voltage (V) increases if the resistance (R) remains constant.

Equal ratios Equivalent fractions are equal ratios. A fraction like $^6/_8$ can be simplified to $^3/_4$. The two fractions, $^3/_4$ and $^6/_8$, are equal ratios. Equal ratios are proportions.

Grade The slope of an incline, or the ratio of vertical rise to horizontal distance covered.

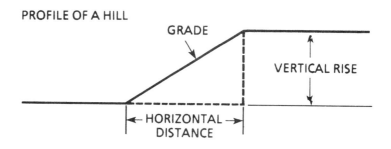

PROFILE OF A HILL

Other similar grade situations are encountered in building roofs, in surveying, and in plumbing.

Inverse proportional relationship A relationship between variables like pressure and volume of a gas ($P \times V =$ Constant) where one variable increases when the other variable decreases.

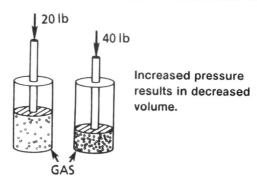

Increased pressure results in decreased volume.

Using Ratios and Proportions

Proportion

A proportion is an expression of equality between two ratios. If a proportion equates equivalent fractions, the fractions can be simplified as shown below.

$$\frac{6}{9} = \frac{4}{6} \; ; \quad \frac{6 \div 3}{9 \div 3} = \frac{4 \div 2}{6 \div 2} \quad \text{or} \quad \frac{2}{3} = \frac{2}{3}$$

Proportions are seldom simplified except to prove that the two ratios are indeed equal to each other.

Rate

A rate is a ratio that usually involves the units of time in the denominator. Examples are 20 miles/hour, 16 feet/second, and 500 gallons/minute.

Ratio

A comparison of two quantities—either pure numbers or numbers with units. Comparisons of quantities with similar units are often called proper ratios. The quantities that are compared need not have similar units.

Similar figures

Figures are **similar** if their corresponding sides are proportional and their corresponding angles are equal. In a general sense, similar figures are overall reductions or enlargements of one another. In the two similar figures shown below, the corresponding pairs of sides are AB and DE, AC and DF, BC and EF. The corresponding angles are CAB and FDE, ABC and DEF, and ACB and DFE.

Proportions can be written between corresponding sides as follows:

$$\frac{AB}{DE} = \frac{AC}{DF} \quad \text{or} \quad \frac{AB}{DE} = \frac{BC}{EF}$$